Probabilities

Probabilities

The Little Numbers that Rule Our Lives

Peter Olofsson

Trinity University
San Antonio, Texas

A John Wiley & Sons, Inc., Publication

Published by John Wiley & Sons, Inc., Hoboken, New Jersey.
Published simultaneously in Canada.

Library of Congress Cataloging-in-Publication Data is available.

ISBN 978-0-470-62445-6

Printed in the United States of America.

10 9 8 7 6 5 4 3 2 1

Preface

This book is about those little numbers that we just cannot escape. Try to remember the last day you didn't hear at least something about probabilities, chance, odds, randomness, risk, or uncertainty. I bet it's been a while. In this book, I will tell you about the mathematics of such things and how it can be used to better understand the world around you. It is not a textbook though. It does not have little colored boxes with definition or theorems, nor does it contain sections with exercises for you to solve. My main purpose is to entertain you, but it is inevitable that you will also learn a thing or two. There are even a few exercises for you, but they are so subtly presented that you might not even notice until you have actually solved them.

The spousal thanks is always more than a formality. I thank Αλκμήνη for putting up with irregular work hours and everything else that comes with writing a book, but also for help with Greek words and for reminding me of some of my old travel stories that you will find in the book. I am deeply grateful to Professor Olle Häggström at Chalmers University of Technology in Göteborg, Sweden. He has read the entire manuscript, and his comments are always insightful, accurate, and clinically free from unnecessary politeness. If you find something in this book that strikes you as particularly silly, chances are that Mr. Häggström has already pointed it out to me but that I decided to keep it for spite. I have also received helpful comments from John Haigh at the

University of Sussex, Steve Quigley at Wiley, and from an anonymous referee. Thanks also to Kris Parrish and Susanne Steitz at Wiley, to Sheree Van Vreede at Sheree Van Vreede Publications Services for excellent copyediting, and to Amy Hendrickson at Texnology Inc. for promptly and patiently answering my LaTeX questions.

A large portion of this book was written during the tumultuous Fall of 2005. Our move from Houston to New Orleans in early August turned out to be a masterpiece of bad timing as Hurricane Katrina hit three weeks later. We evacuated to Houston, and when Katrina's sister Rita approached, we took refuge in the deserts of West Texas and New Mexico. Sandstorms are so much more pleasant than hurricanes! However, it was also nice to return to New Orleans in January 2006; the city is still beautiful, and its chargrilled oysters are unsurpassed. I am grateful to many people who housed us and helped us in various ways during the Fall and by doing so had direct or indirect impact on this book. Special thanks to Kathy Ensor & Co. at the Department of Statistics at Rice University in Houston and to Tom English & Co. at the College of the Mainland in Texas City for providing me with office space. Finally, thanks to Professor Peter Jagers at Chalmers University of Technology, who as my Ph.D. thesis advisor once in a distant past wisely guided me through my first serious encounters with probabilities, those little numbers that rule our lives.

PETER OLOFSSON
www.peterolofsson.com

New Orleans, 2006

Contents

1

Computing Probabilities: Right Ways and Wrong Ways

THE PROBABILIST

Whether you like it or not, probabilities rule your life. If you have ever tried to make a living as a gambler, you are painfully aware of this, but even those of us with more mundane life stories are constantly affected by these little numbers. Some examples from daily life where probability calculations are involved are the determination of insurance premiums, the introduction of new medications on the market, opinion polls, weather forecasts, and DNA evidence in courts. Probabilities also rule who you are. Did daddy pass you the X or the Y chromosome? Did you inherit grandma's big nose? And on a more profound level, quantum physicists teach us that *everything* is governed by the laws of probability. They toss around terms like the *Schrödinger wave equation* and *Heisenberg's uncertainty principle,* which are much too difficult for most of us to understand, but one thing they do mean is that the fundamental laws of physics can only be stated in terms of probabilities. And the fact that Newton's deterministic laws of physics are still useful can also be attributed to results from the theory of probabilities. Meanwhile, in everyday life, many of us use probabilities in our language and say things like "I'm 99% certain" or "There is a one-in-a-million chance" or, when something unusual happens, ask the rhetorical question "What are the odds?"

Some of us make a living from probabilities, by developing new theory and finding new applications, by teaching others how to use them, and occasionally by writing books about them. We call ourselves *probabilists*. In the universities, you find us in mathematics and statistics departments; there are no departments of probability. The terms "mathematician" and "statistician" are much more well known than "probabilist," and we are a little bit of both but we don't always like to admit it. If I introduce myself as a mathematician at a cocktail party, people wish they could walk away. If I introduce myself as a statistician, they do. If I introduce myself as a probabilist...well, most actually still walk away. They get upset that somebody who sounds like the Swedish Chef from the Muppet Show tries to impress them with difficult words. But some stay and give me the opportunity to tell them some of the things I will now tell you about.

Let us be etymologists for a while and start with the word itself, *probability*. The Latin roots are *probare*, which means to test, to prove, or to approve, and *habilis*, which means apt, skillful, able. The word "probable" was originally used in the sense "worthy of approval," and its connection to randomness came later when it came to mean "likely" or "reasonable." In my native Swedish, the word for probable is "sannolik," which literally means "truthlike" as does the German word "wahrscheinlich." The word "probability" still has room for nuances in the English language, and Merriam-Webster's online dictionary lists four slightly different meanings. To us a probability is a number used to describe how likely something is to occur, and probability (without indefinite article) is the study of probabilities.

Probabilities are used in situations that involve randomness. Many clever people have thought about and debated what randomness really is, and we could get into a long philosophical discussion that could fill the rest of the book. Let's not. The French mathematician Pierre-Simon Laplace (1749–1827) put it nicely: "Probability is composed partly of our ignorance, partly of our knowledge." Inspired by Monsieur Laplace, let us agree that you can use probabilities whenever you are faced with uncertainty. You could:

- Toss a coin, roll a die, spin a roulette wheel
- Watch the stock market, the weather, the Super Bowl
- Wonder if there is an oil well in your backyard, if there is life on Mars, if Elvis is alive

These examples differ from each other. The first three are cases where the outcomes are equally likely. Each individual outcome has a probability that is simply one divided by the number of outcomes. The probability is $1/2$

to toss heads, 1/6 to roll a 6, and 1/38 to get the number 29 in roulette (an American roulette wheel has the numbers 1–36, 0, and 00). Pure and simple. We can also compute probabilities of groups of outcomes. For example, what is the probability to get an odd number when rolling a die? As there are three odd outcomes out of six total, the answer is $3/6 = 1/2$. These are examples of *classical probability*, the first type of probability problems studied by mathematicians, most notably, Frenchmen Pierre de Fermat and Blaise Pascal whose seventeenth century correspondence with each other is usually considered to have started the systematic study of probabilities. You will learn more about Fermat and Pascal later in the book.

The next three examples are cases where we must use data to be able to assign probabilities. If it has been observed that under current weather conditions it has rained about 20% of the days, we can say that the probability of rain today is 20%. This probability may change as more weather data are gathered and we can call it a *statistical probability*. As for the 2006 Super Bowl, I placed a bet on the Houston Texans that gave odds of 800 to 1, which means that the bookmaker assigned a probability of less than 1/800 that the Texans would win. However he came to this conclusion, he must have used plenty of data other than that he once spent a summer in Houston and almost died of heatstroke.

The third trio of examples is different from the previous two in the sense that the outcome is already fixed; you just don't know what it is. Either there is an oil well or there isn't. Before you start drilling, you still want to have some idea of how likely you are to find oil and a geologist might tell you that the probability is about 75%. This percentage does not mean that the oil well is there nine months of the year and slides over to your neighbor the other three, but it does mean that the geologist thinks that your chances are pretty good. Another geologist may tell you the probability is 85%, which is a different number but means the same thing: Chances are pretty good. We call these *subjective probabilities*. In the case of a living Elvis, I suppose that depending on whom you ask you would get either 0% or 100%. I mean, who would say 25%? Little Richard?

Some knowledge about proportions may be helpful when assigning subjective probabilities. For example, suppose that your Aunt Jane in Pittsburgh calls and tells you that her new neighbor seems nice and has a job that "has something to do with the stars, astrologer or astronomer." Without having more information, what is the probability that the neighbor is an astronomer? As you have virtually no information, would you say 50%? Some people might. But you should really take into account that there are about four times as many astrologers as astronomers in the United States, so a probability of 20% is more realistic. Just because something is "either/or" does not mean it is "50–50." Andy Rooney may have been more insightful than he intended when he stated his 50–50–90 rule: "Anytime you have a 50–50 chance of getting something right, there's a 90% probability you'll get it wrong."

THE PROBABILIST'S TOYS AND LANGUAGE

Probabilists love to play with coins and dice. In a Platonic sense. We like the *idea* of tossing coins and rolling dice as experiments that have equally likely outcomes. Suppose that a family with four children is chosen at random. What is the probability that all four are girls? A coin-tossing analogy would be to ask for the probability to get four heads when a coin is tossed four times. Many probability problems can be illustrated by coin tossing, but this would quickly become boring so we introduce variation by also rolling dice, spinning roulette wheels, picking balls from urns, or drawing from decks of cards. Dice, roulette, and card games are also interesting in their own right, and you will find a chapter on gambling later in the book. Of course. Probability without gambling is like beer without bubbles.

Probability is the art of being certain of how uncertain you are. The statement "the probability to get heads is $1/2$" is a precise statement. It tells you that you are as likely to get heads as you are to get tails. Another way to think about probabilities is in terms of average long-term behavior. In this case, if you toss the coin repeatedly, in the long run you will get roughly 50% heads and 50% tails. Of this you can be certain. What you cannot be certain of is how the next toss will come up.

Probabilists use special terminology. For example, we often refer to a situation where there is uncertainty as an "experiment." This situation could be an actual experiment such as tossing a coin or rolling a die, but also something completely different such as following the stock market or watching the Wimbledon final. An experiment results in an *outcome* such as "heads," "6," "Volvo went up," or "Björn Borg won" (those were the days). A group of outcomes is called an *event*. In plain language, an event is something that can happen in an experiment. It can be a single outcome (roll 6) or a group of outcomes (roll an odd number). The mathematical description of an event is that it is a *subset* of the set of all possible outcomes, and mathematicians would describe outcomes as *elements* of this set. Probabilists use the words

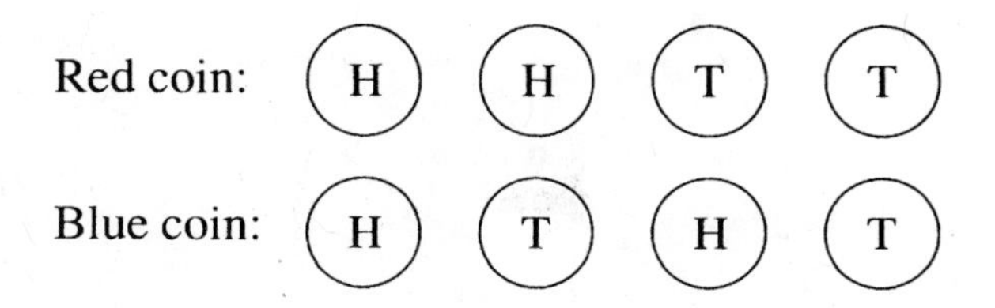

Figure 1.1 The four equally likely outcomes when you toss two coins.

"outcome" and "event" to emphasize the connection with things that happen in reality. In formulas, we denote events by uppercase letters and use the letter "P" to denote probability. The mathematical expression P(A) should thus be read "the probability of (the event) A." We may also talk about the probability of a statement rather than an event. However, it is mere language; the verbal description of an event is of course a statement.

The set of all possible outcomes is called the *sample space.*[1] Sometimes there is more than one choice of sample space. For example, suppose that you toss two coins and ask for the probability that you get two heads. As the number of heads can be 0, 1, or 2, you might be tempted to take these three numbers as the sample space and conclude that the probability to get two heads is $1/3$. However, if you repeated this experiment, you would notice after a while that you tend to get two heads less than one third of the tosses. The problem is that your sample space consists of three outcomes that are *not equally likely*. Let us distinguish between the two coins by painting one red and the other blue. There are then four equally likely outcomes: both show heads; the red shows heads and the blue shows tails; the red shows tails and the blue shows heads; and both show tails. In a more convenient notation, our sample space consists of the four equally likely outcomes HH, HT, TH, and TT. One out of four gives two heads, and the correct probability is $1/4$. See Figure 1.1 for an illustration of the four equally likely outcomes.

Here is a similar problem. If you roll two dice, what is the probability that the sum of the two equals eight? First note that the sum of two dice can be any of the numbers 2, 3, ..., 12 but that these are not equally likely. To find the equally likely outcomes, we need to distinguish between the two dice, for

[1]The term "sample space" was coined by mathematician and Austro-Hungarian fighter pilot Richard von Mises. That is, he coined the German term *Merkmahlraum* (label space), which appears in his 1931 book with the impressive German title *Wahrscheinlichkeitsrechnung* (probability calculus).

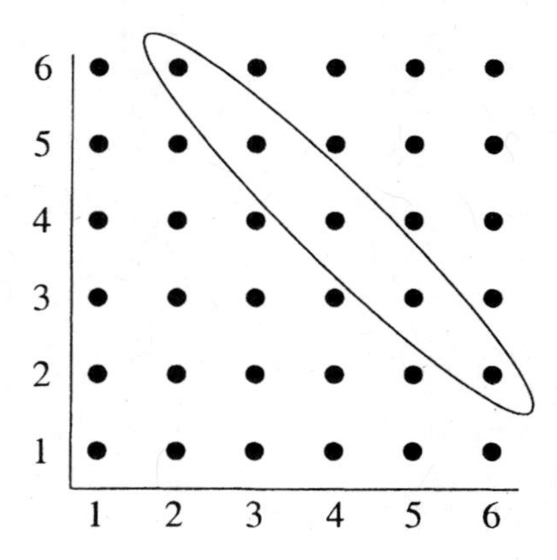

Figure 1.2 The sample space of 36 equally likely outcomes when you roll two dice. The event that the sum equals eight is marked; note that it consists of five outcomes because there are two ways to get 2 and 6 as well as 3 and 5 but only one way to get 4 and 4.

example, by pretending that they have different colors, red and blue, just like we did with the two coins above, and consider 36 possible outcomes. As the sum can be eight by adding $2+6$, $3+5$, or $4+4$, we might first think that there are 3 possibilities out of 36 to get sum eight, but we also need to distinguish, for example, between the cases "blue die equals 2 and red die equals 6" on the one hand and "blue die equals 6 and red die equals 2" on the other. If we make this distinction, we realize that there are five ways to get sum eight and the probability is $5/36$. See Figure 1.2 for an illustration of the sample space of 36 equally likely outcomes and the event that the sum equals eight.

Here is another example of a similar nature. Consider a randomly chosen family with three children. What is the probability that they have exactly one daughter? There can be 0, 1, 2, or 3 girls, but you know by now that these are not equally likely. Instead, distinguish the kids by birth order so that, for example, BGB means that the first child is a boy, the second a girl, and the third a boy. The eight equally likely outcomes are as follows:

BBB, BBG, BGB, GBB, BGG, GBG, GGB, GGG

We're on easy street now; just note that three of the eight outcomes have one girl, and the probability of exactly one girl is therefore $3/8$. Now consider a randomly chosen girl who has two siblings. What is the probability that she has no sisters? This situation looks similar. If she has no sisters, this means that her family has three children, exactly one of whom is a girl and we just saw that the probability of this is $3/8$. Convinced? You should not be. This situation is different because we are not choosing a *family* with three children; we are choosing a *girl* who belongs to such a family. Thus, the outcome BBB is impossible. Is the probability then $3/7$? Think about this for a while before you read on.

I hope you answered no. We need a completely new sample space that also accounts for the chosen girl. If we denote her by an asterisk, the 12 equally likely outcomes are as follows:

$$BBG^*,\ BG^*B,\ G^*BB,\ BG^*G,\ BGG^*,\ G^*BG$$

$$GBG^*,\ G^*GB,\ GG^*B,\ G^*GG,\ GG^*G,\ GGG^*$$

and the probability that she has no sisters is $3/12 = 1/4$. Note how the previous outcomes are now split up according to how many girls they contain. The one with three girls, GGG, is split up into three equally likely outcomes because either of the three girls may be the chosen one. The probabilities that we have computed show that 37.5% of three-children families have exactly one daughter and 25% of girls from three-children families have no sisters.

What is the probability that all three children are of the same gender? Consider the following *faulty* argument: Two children must always be of the same gender. Whatever this gender is, the third child is equally likely to be of this gender or not, and thus the probability that all three are of the same gender is $1/2$. This example is a variant of a coin-tossing problem given by the British nobleman and amateur scientist Sir Francis Galton (about whom you will learn more in chapters to come) in 1894 to illustrate the dangers of sloppy thinking. Use our first sample space to discover the error, and argue that the correct probability is $1/4$.

Let us next consider an old gambling problem that goes along the same lines. I have three dice and offer you even odds to play the following game: The dice are rolled, and their sum is computed. If the sum is nine, you win. If it is ten, I win. If it is neither, I roll again. Is this game fair?

There are six ways in which the sum can be nine:

$$1+2+6,\quad 1+3+5,\quad 1+4+4,\quad 2+2+5,\quad 2+3+4,\quad 3+3+3$$

and likewise there are six ways to get sum ten:

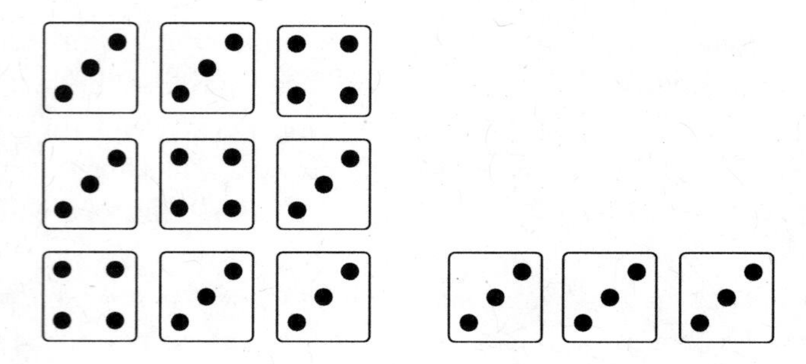

Figure 1.3 Three ways to get sum ten from two 3s and a 4 (left), but only one way to get sum nine from three 3s (right).

$$1+3+6, \quad 1+4+5, \quad 2+2+6, \quad 2+3+5, \quad 2+4+4, \quad 3+3+4$$

It sure looks like the game is fair, but beware, in the long run, I would slowly but surely win your money. But why?

Before you decide to play, you need to first identify the equally likely outcomes. And just like in the case of the two dice earlier, it is helpful to imagine that the three dice have three different colors, for example, red, green, and blue. If we list the dice in this order, the equally likely outcomes are (1,1,1), (1,1,2), (1,2,1), (2,1,1), (2,2,1), and so on until (6,6,6); a moment's thought reveals that there are $6 \times 6 \times 6 = 216$ of them. Let us look at one of the ways to get sum nine, $1 + 4 + 4$. This sum corresponds to three of the equally likely outcomes: (1,4,4), (4,1,4), and (4,4,1). If we instead consider $1 + 2 + 6$, this corresponds to six outcomes: (1,2,6), (1,6,2), (2,1,6), (2,6,1), (6,1,2), and (6,2,1). In general, if all three dice show different numbers, this can occur in six ways; if two show the same number, this can occur in three ways; and if all three are the same, this can only occur in one way.

Now count above to realize that 27 outcomes give sum ten and only 25 give sum nine. The tie-breaker is the last outcome: There is only one way to combine $3 + 3 + 3$ but three ways to combine $3 + 3 + 4$; see Figure 1.3 for an illustration. Thus, out of the 52 outcomes that give a winner, I win in 27, or about 52%, and you win in the remaining 25, or 48%. Not a big difference, but it would be enough to make a living (some venture capital needed).

I mentioned that this problem is an old one. It was in fact solved almost 400 years ago by the great astronomer and telescope builder Galileo after being approached by a group of gambling Florentine noblemen. It is amusing to imagine how the world's most brilliant scientist of his time spent time helping people with their gambling problems. Good thing for Einstein that there were no casinos in Atlantic City in the 1930s; his Princeton office might have been flooded by gamblers having spent the last of their money on a bus ticket, desperate for help from the genius.

We are often interested in more than one event. For example, suppose that people are chosen for an opinion poll and asked about their smoking habits and political sympathies. Consider one selected person. Let us denote the event that she is a smoker by S and the event that she is a Republican by R. We can then make up new events. The event that she is a smoker *and* a Republican is a new event, which we write as "S and R." The event that she is a smoker *or* a Republican is another new event, written as "S or R." It is important to know that we by "S or R" mean "smoker or Republican *or both*." This definition of "or" is typical in mathematics, logic, and computer science. In daily language, it is often emphasized by using the expression "and/or" to distinguish from what math people call the *exclusive or*, which only permits one of the two, like in the phrase "You want fries or onion rings with that?"

The event that the selected individual is not a Republican is simply written as "not R." The event that she is neither a Republican nor a smoker can be expressed in two different ways. One way is to negate that she is either, which gives "not (R or S)." The other way is to negate each separately and put them together: "(not R) and (not S)." We have argued for the following equality between events:

$$\text{not (R or S)} = \text{(not R) and (not S)}$$

The parentheses are there to make it clear to what "not" refers. In a similar way,

$$\text{not (R and S)} = \text{(not R) or (not S)}$$

Make sure that you understand these little exercises in logic; we will make use of them later.

THE PROBABILIST'S RULE BOOK

Probabilities can be expressed as fractions, as decimal numbers, or as percentages. If you toss a coin, the probability to get heads is $1/2$, which is the same as 0.5, which is the same as 50%. There are no rules for when to use which notation, and you will see examples of all three in this book. In daily language, proper fractions are often used and often expressed, for example, as "one in ten" instead of 1/10 ("one tenth"). This is also natural when you deal with equally likely outcomes. Decimal numbers are more common in technical and scientific reporting when probabilities are calculated from data. Percentages are also common in daily language and often with "chance" replacing "probability." Meteorologists, for example, typically say things like "there is a 20% chance of rain." The phrase "the probability of rain is 0.2" means the same thing. When we deal with probabilities from a theoretical viewpoint, we always think of them as numbers between 0 and 1, not as percentages.

Regardless of how probabilities are expressed, they must follow certain rules. One such rule that is easy to understand is that a probability can never be a negative number. The lowest possible probability is 0, meaning that we are dealing with something that just does not happen. There is no point in trying to emphasize this further by letting the probability be −0.3 or −5.[2] A related rule is that a probability can never be more than 1 (or 100%). If the probability is 1 (or 100%), we are describing something that we are absolutely certain about. Of course you can still say that you are 200% certain that the Texas Rangers will win the World Series, but nobody outside Dallas will take you seriously.

The next rule is that the probability that something does *not* occur can be computed as one minus the probability that it does occur. In a formula,

$$\text{P(not A)} = 1 - \text{P(A)}$$

Also easy to accept. The probability not to get 6 when you roll a die is 5/6, which is also equal to $1-1/6$. If the chance of rain is 20%, then the chance that it does not rain is 80%. In all its simplicity, this rule turns out to be surprisingly useful. In fact, in his excellent book *Taking Chances: Winning with Probability*, British probabilist John Haigh names it probability's Trick Number One.

In the world of gambling, probabilities are often expressed by *odds*. To say that the odds are 4:1 against the event A means that it is four times as likely that A does not occur than that it occurs. We get the equation $\text{P(not A)} = 4 \times \text{P(A)}$,

[2] I do not know how familiar you are with negative numbers, but to mathematicians they are as natural as air and water. Here is the world's funniest math joke: A biologist, a physicist, and a mathematician are sitting at a sidewalk cafe watching a house across the street. After a while two people enter the house. A little later, three people exit. "Reproduction," says the biologist. "Measurement error," says the physicist. "Hmm," says the mathematician, "if a person enters the house it will be empty again."

which has the solution $P(A) = 1/5$ and $P(\text{not } A) = 4/5$. As bookmakers are in the business to make a living, offering odds of 4:1 in reality means that they think that the probability of A is *less* than 1/5.

Another rule. Let A and B be events such that whenever A occurs, B must also occur. Then P(A) is less than (or equal to) P(B), and the mathematical notation for this is $P(A) \leq P(B)$. For an example, let A be the event to roll a 6 and B the event to roll an even number. Whenever A occurs, B must also occur. However, B can occur without A occurring if you roll 2 or 4. In particular, the composition of two events is always less probable than each individual event. What I mean is that P(A and B) is always less than both P(A) and P(B), regardless of what A and B are.

As an example of the rule from the last paragraph, let us consider Mrs. Boudreaux and Mrs. Thibodeaux who are chatting over their fence when the new neighbor walks by. He is a man in his sixties with shabby clothes and a distinct smell of cheap whiskey. Mrs. B, who has seen him before, tells Mrs. T that he is a former Louisiana state senator. Mrs. T finds this very hard to believe. "Yes," says Mrs. B, "he is a former state senator who got into a scandal long ago, had to resign, and started drinking." "Oh," says Mrs. T, "that sounds more likely." "No," says Mrs. B, "I think you mean less likely."

Strictly speaking, Mrs. B is right. Consider the following two statements about the shabby man: "He is a former state senator" and "He is a former state senator who got into a scandal long ago, had to resign, and started drinking." It is tempting to think that the second is more likely because it gives a more exhaustive explanation of the situation at hand. However, this reason is precisely why it is a *less* likely statement. Note that whenever somebody satisfies the second description, he must also satisfy the first but not vice versa. Thus, the second statement has a lower probability (from Mrs. T's subjective point of view; Mrs. B of course knows who the man is). This example is a variant of examples presented in the book *Judgment under Uncertainty* by Economics Nobel laureate[3] Daniel Kahneman and co-authors Paul Slovic and Amos Tversky. They show empirically how people often make similar mistakes when they are asked to choose the most probable among a set of statements. It certainly helps to know the rules of probability. A more discomforting aspect

[3]But I want to point out that the Economics prize is not a "true" Nobel prize in the sense that it was not mentioned in Alfred Nobel's will. The prize was first awarded in 1969, and its official name is "The Bank of Sweden Prize in Economic Sciences in Memory of Alfred Nobel." Just so that you know.

is that the more you explain something in detail, the more likely you are to be wrong. If you want to be credible, be vague.

The final rule is the *addition rule*. It says that in order to get the probability that either of two events occur, you add the probabilities of the two individual events. This rule, however, only applies if the two events in question *cannot occur at the same time* (the technical term for such events is that they are *mutually exclusive*). In a formula:

$$P(A \text{ or } B) = P(A) + P(B)$$

For example, roll a die and consider the events A: to get 6 and B: to get an odd number. These events qualify as mutually exclusive because you cannot get both 6 and an odd number in the same roll. It is "same roll" that is important here; of course you can get 6 in one roll and an odd number in the next. By the formula above, the probability to get 6 or an odd number in the same roll is $1/6 + 3/6 = 4/6$.

In his bestseller *Innumeracy*, John Allen Paulos tells the story of how he once heard a local weatherman claim that there was a 50% chance of rain on Saturday and a 50% chance of rain on Sunday and thus a 100% chance of rain during the weekend. Clearly absurd, but what is the error? Faulty use of the addition rule! As a rainy Saturday does not exclude a rainy Sunday, we here have two events that can both occur the same weekend. In cases like this one, there is a modified version of the addition rule that says that you first add the two probabilities as before and then subtract the probability that *both* events occur. In a formula, it looks as follows:

$$P(A \text{ or } B) = P(A) + P(B) - P(A \text{ and } B)$$

Note that if A and B cannot occur at the same time, then $P(A \text{ and } B) = 0$ and we have the first addition rule as a special case. If we let A denote the event that it rains on Saturday and B the event that it rains on Sunday, the event "A and B" describes the case in which it rains both days. To get the probability of rain over the weekend, we now add 50% and 50%, which gives 100%, but we must then subtract the probability that it rains both days. Whatever this is, it is certainly more than 0 so we end up with something less than 100%, just like common sense tells us that we should. I just wonder what the weatherman would have said if the chances of rain had been 75% each day.

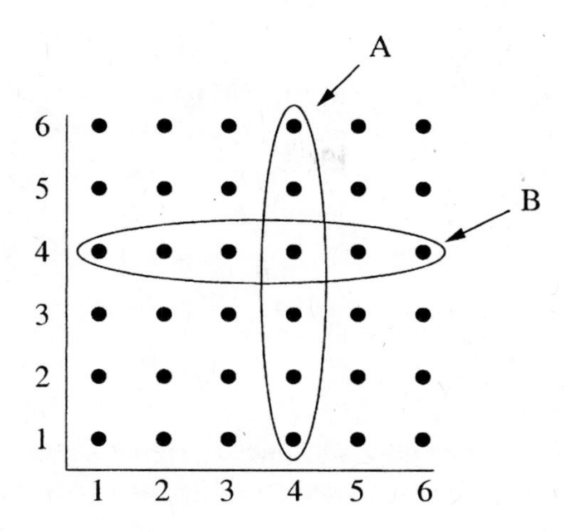

Figure 1.4 The sample space of 36 equally likely outcomes for rolling two dice. The events "4 on first die" and "4 on second die" are marked, and you may note that there are 6 outcomes in each event, 11 outcomes that are in at least one event, and 1 outcome that is in both.

Let us also check the formula in a dice example. If you roll two dice, what is the probability to get at least one 4? Here, the relevant events are A: 4 on the first die and B: 4 on the second die. The event to get at least one 4 is then the event "A or B," and in Figure 1.4, you can check directly that this equals 11/36. Also, P(A) = 6/36, P(B) = 6/36, and P(A and B) = 1/36 because there is only one outcome that gives 4 on both dice. As $6 + 6 - 1 = 11$, the formula is valid.

Whenever probabilities are assigned, this must be done in a way such that none of the rules are violated. Ask a friend how likely he thinks it is that it will rain Saturday, Sunday, both days, and at least one of the days, respectively. You will then get four probabilities that must satisfy the rules that we have discussed above. For example, somebody may think that rain on Saturday is pretty likely, say 70%, and the same for Sunday. Rain both days? Well, maybe 50%. For the last probability, let's say 80%. But this assignment of probabilities violates the addition rule because 80 is not equal to $70 + 70 - 50 = 90$. Somebody else might come up with the following probabilities (same order): 70%, 60%, 80%, and 50%. These do satisfy the addition rule but suffer from another problem. Can you tell which? (Hint: Mrs. Boudreaux could.)

Let us keep thinking about weekend weather. Suppose that both Saturday and Sunday each have probability 0.5 to get rain and that the probability is p

that it rains both days (we now think of probabilities as numbers between 0 and 1, not percentages). What is the range of possible values of p? How does the probability of rain during the weekend depend on p?

If we let A and B be the events "rain on Saturday" and "rain on Sunday" respectively, then a rainy weekend is the event "A or B," and because $p = P(A \text{ and } B)$, we get the equation

$$P(A \text{ or } B) = P(A) + P(B) - P(A \text{ and } B) = 1 - p$$

As p must be less than both $P(A)$ and $P(B)$, it cannot be more than 0.5. If p is 0, then $P(A \text{ or } B) = 1$ and the rainy weekend is a fact. As p ranges from 0 to 0.5, the probability of a rainy weekend decreases from 1 to 0.5. Why? It has to do with how likely rainy Saturdays and Sundays are to come in pairs. Think of a year, which has 52 weekends. On average, we expect to get rain 26 Saturdays and 26 Sundays. If p is 0, this means that if it rains on a Saturday, it *never* rains on the following Sunday and if it does not rain on Saturday, it *always* rains on Sunday. Thus, the 26 rainy Saturdays and 26 rainy Sundays must be spread over the year so that they never come in pairs. The only way to do this is to let every weekend have exactly one rainy day. As p gets bigger, rainy days are more likely to come in pairs, and the extreme case is when $p = 0.5$. Then *all* rainy days come in pairs and the year has half of its weekends rainy and the other half dry.

Here is an exercise for you. Change the probabilities a little, and let $P(A) = 0.6$ and $P(B) = 0.7$, and let p again denote $P(A \text{ and } B)$. Explain why p must be between 0.3 and 0.6.

INDEPENDENCE, AIRPLANES, AND RUSSIAN PEASANTS

Plenty of random things happen in the world all the time, most of which have nothing to do with one another. If you toss a coin and I roll a die, the probability that you get heads is $1/2$ regardless of the outcome of my die. If there is a 20% chance of rain tomorrow, this does not change if a flu outbreak in Asia is reported. Changes in the U.S. stock market indexes have nothing to do with who wins the Wimbledon tennis tournament. Events that in this way are unrelated to each other are called *independent*. It is easy to compute the probability that two independent events both occur; simply multiply the probabilities of the two events. We call this computation the *multiplication rule* for probabilities, described in a formula as

$$P(\text{A and B}) = P(A) \times P(B)$$

It works in two directions. If we can argue that two events are independent, then we can use the multiplication rule to compute the probability that both occur at the same time. Conversely, if we can show that the multiplication rule holds, then we can conclude that the events are independent. It can be argued at some length why this is true and we will just look at some simple examples to convince ourselves that formula and intuition agree. Let us do the first example above, that you toss a coin and I roll a die. There are 12 equally likely outcomes: (H,1), ..., (H,6), (T,1), ..., (T,6) in the obvious notation. What is now the probability that you toss heads and I roll a 6? Obviously $1/12$. The individual probabilities of heads and 6 are $1/2$ and $1/6$, respectively, and $1/2 \times 1/6$ equals $1/12$ indeed.

For another example, take a deck of cards, draw one card, and consider the two events, A: to get an ace, and H: to get hearts. Are these independent? Let us check whether the multiplication rule holds. The individual probabilities are

$$P(A) = 4/52 = 1/13$$
$$P(H) = 13/52 = 1/4$$

and the probability to get both A and H is the probability to get the ace of hearts, which is $1/52$, which is the product of $1/13$ and $1/4$. We have

$$P(\text{A and H}) = P(A) \times P(H)$$

which means that A and H are independent. Now remove the two of spades from the deck, reshuffle, and consider the same two events as above. Are they still independent? They must be, right? After all, the two of spades has nothing to do with either aces or hearts. Let us compute the probabilities. There are now 51 cards, and we get

$$P(A) = 4/51$$
$$P(H) = 13/51$$

and $P(\text{A and H}) = 1/51$. As $P(\text{A and H})$ is not equal to $P(A) \times P(H)$, we must conclude that the events are not independent anymore. What happened?

Removing the two of spades changes the proportions of aces in the deck from $4/52$ to $44/51$, but not *within the suit of hearts* where it remains at $1/13 = 4/52$. Here is how you should think about independent events: *If one event has occurred, the probability of the other does not change.* In the card example, the probability of A is $4/51$ but changes to $1/13$ if the event H occurs.

Here is a question I often ask my students after I have introduced independence: If two events cannot occur at the same time, are they independent? At first you might think so. After all, they have nothing to do with each other, right? Wrong! They have a lot to do with each other. If one has occurred, we know for certain that the other cannot occur. The probability to roll a 6 is $1/6$, but if I tell you that the outcome is an odd number, the probability of a 6 drops down to 0. Think this through. It is important to understand independence.

There is a story that is sometimes told about the great Russian mathematician Andrey Nikolaevich Kolmogorov, among many other things the founder of the modern theory of probability. In Stalin's Soviet Union in the 1930s, the concept of independence did not fit well with the historical determinism of Marxist ideology. When questioned by a panel of ideologues about this possible heresy, Kolmogorov countered, "If the peasants pray for rain and it actually starts to rain, were their prayers answered?" The atheist ideologues had to confess that this must indeed be a case of independent events and Kolmogorov lived a long and productive life until his death in 1987 at the age of 84.

In December 1992, a small passenger airplane crashed in a residential neighborhood near Bromma airport outside Stockholm in Sweden, causing no death or injury to any of the residents. Already disturbed by increasing traffic and expansion plans for the airport, the residents now got more reasons to worry. In an effort to calm people, the airport manager said in an interview on TV that statistically people should now feel safer because the probability to have another accident had become so much smaller than before. I was at the time a graduate student in Sweden, studying probability and statistics, and thought that it was amusing to hear both "statistically" and "probability" used in the same sentence in such a careless way. In youthful vigor, I immediately wrote a letter that was published in some leading Swedish newspapers, where I explained why the airport manager's statement was incorrect. I also encouraged him to contact me so that I could recommend a good probability textbook. I never heard from him.

The airport manager's error is common: He confuses the probability that something happens *twice* and the probability that something happens *again*. Toss a coin twice. What is the probability to get heads twice? One fourth. Toss a coin until you get heads. What is the probability that you get heads again in the next toss? One half, by independence. Replace the coin tosses with flights to and from Bromma Airport and the probability of tossing heads with the probability of having a crash, and you got him. His only possible defense would be that crashes are not independent, and that after such a crash, an investigation is started that may improve security. Perhaps. But first of all, that was not his argument. He believed that there was magic in the sheer probabilities. Second, even if there was such an investigation, it would not be likely to dramatically reduce the probability of another crash, which can occur for many different reasons. The events are not independent, but almost. Compare with the example above where the events "ace" and "hearts" are not independent when a card is drawn from a deck without the two of spades. The probability to get an ace is $4/51$, which is roughly 0.078, and the probability to get an ace if we know that the card is hearts is $1/13$, or roughly 0.077, not much different. The events are almost independent.

In a probability class, I once pointed out that even if you have just tossed nine heads in a row, the next toss is still equally likely to give another head as it is to give tails. A student approached me after class and wondered how this could be possible. After all, aren't sequences of ten consecutive heads pretty rare? The first reply is that a coin has no memory. When you start tossing a coin, would you need to know whether the coin has been tossed before and what it gave? Of course not. The student had no problem accepting this assertion but still insisted that if he was to toss a coin repeatedly, sequences of ten consecutive heads would be very rare, which would contradict my claim. Although he is right that a sequence of ten consecutive heads is pretty rare (it has probability $1/1{,}024$, less than one thousandth), this is irrelevant because I was talking about the probability to get heads once more after we had already gotten nine in a row. If he tossed his coin repeatedly in sequences of ten, he would start with nine consecutive heads about once every 512 times and about half of these would finish with yet another head in the tenth toss. Probability of ten consecutive heads: $1/1{,}024$, probability of heads once more after nine consecutive heads to start with: $1/2$. Airport managers and college students are not alone. These types of mistakes are very common, and I will address them in more depth and detail in later chapters.

Suppose now that you have agreed to settle a dispute with cousin Joe by tossing a coin. The problem is that neither of you has any change. Joe suggests that you instead toss a bottle cap, which will count as heads if it lands with the top up, and tails otherwise. As you cannot assume that these are equally likely, is there any way in which fairness can be guaranteed?

You can suggest a trick invented by computer pioneer John von Neumann. Instead of tossing the cap once and observing heads or tails, the cap is tossed twice. If this gives the sequence HT, you win; if it gives TH, Joe wins. If it gives HH or TT, nobody wins and you start over. Suppose the the probability of heads is some value p, not necessarily $1/2$. As the probability of tails is then $1 - p$, independence gives that the probability to get HT is $p \times (1 - p)$ and the probability to get TH is $(1 - p) \times p$, the same. The procedure is fair (but may take a while if p is very close to 0 or 1).

For independence of more than two events, the multiplication rule still applies. If A, B, and C are independent, then $\mathrm{P(A \text{ and } B)} = \mathrm{P(A)} \times \mathrm{P(B)}$, and similar for the combinations A–C and B–C. Also, the probability that all three events occur is $\mathrm{P(A \text{ and } B \text{ and } C)} = \mathrm{P(A)} \times \mathrm{P(B)} \times \mathrm{P(C)}$. Things are a bit more complicated with three events. It is not enough that the events are independent two by two as the following example shows. I will let you do it on your own. You toss a coin twice and consider the three events

A: heads in first toss

B: heads in second toss

C: different in first and second toss

Show that the events are independent two by two but that C is not independent of the event "A and B" and that the multiplication rule fails for all three events. Note that A alone does not give any information about C, and neither does B alone. However, A and B in combination tells us that C cannot occur.

If you want to compute the probability that at least one of several independent events occur, Trick Number One from page 10 comes in handy. First compute the probability that *none* of the events occurs, and then subtract this probability from one. For example, in the carnival game *chuck-a-luck* you roll three dice and win a prize if you get at least one 6. What is the probability that you win? The probability to roll 6 with one die is $1/6$, and as you have three attempts, you might think that you have a 50–50 chance. It is certainly true that three times $1/6$ equals $1/2$, but this is irrelevant to the problem. If you follow the advice I just gave, first compute the probability that none of the dice gives 6. By independence, this probability is

$$P(\text{no 6s}) = 5/6 \times 5/6 \times 5/6 = (5/6)^3$$

and we get

$$P(\text{at least one 6}) = 1 - (5/6)^3 \approx 0.42$$

and, as always in games that somebody wants you to pay money to play, you are more likely to lose than to win. What if there are instead four dice? Your chance to win is then $1 - (5/6)^4$, which is approximately 0.52 so with four dice you would have an edge.

Another example. An American roulette table has the numbers 1–36, plus 0 and 00. Thus, if you bet on a single number, your chance to win is $1/38$. How many rounds do you have to play if you want to have a 50–50 chance to win at least once? Perhaps 19 rounds (half of 38)? Call the number n. By the same argument as above, we get the equation

$$P(\text{win at least once}) = 1 - (37/38)^n$$

For $n = 19$, this is only about 0.4. For $n = 25$, it is approximately equal to 0.49, and for $n = 26$, it is just above 0.5. You need to play 26 rounds. That 38 divided by 2 equals 19 is another example of something that is true but irrelevant.[4] The number 19 arises in a different way though; If you instead bet on 19 different numbers in one round, you have a 50–50 chance to win. Of course, you can then only win once, whereas with successive bets on the same number, you can win many times. As we shall see later, in the long run, you lose just as much regardless of how you play. Unfortunately.

[4]Which reminds me of the marginally funny story about a man in a balloon who is lost and asks a man on the ground where he is. The man replies, "You are in a balloon." "Just my luck," says the balloonist, "asking a mathematician." "How did you know I'm a mathematician?" asks the man, and the balloonist replies, "Your answer was correct but useless!"

CONDITIONAL PROBABILITY, SWEDISH TV, AND BRITISH COURTS

If two events are not independent, they are called...get ready now...dependent. If two events are dependent, the probability of one changes with the knowledge of whether the other has occurred. The probability to roll a 6 is $1/6$. If I tell you that the outcome is an even number, you can rule out the outcomes 1, 3, and 5, and the probability to get 6 changes to $1/3$. We call this the *conditional probability* of getting 6 *given* that the outcome is even. I have mentioned that you can think of probabilities in terms of average long-term behavior. The same is true for conditional probabilities; you just ignore all outcomes that do not satisfy the condition. In the dice example I just gave, you would thus disregard all odd outcomes and count the proportion of 6s among the even outcomes, and this should stay close to one third after a while. There is a multiplication rule that can be stated in terms of conditional probabilities. For any events A and B, the following is always true:

$$\text{P(A and B)} = \text{P(A)} \times \text{P(B given A)}$$

In other words, to find the probability that both A and B occur, first find the probability of A, and then the conditional probability of B given that A occurred. In applications of the formula, it is up to you which event you want to call A and which to call B. Suppose that you draw two cards from a deck. What is the probability that both are aces? The probability that the first card is an ace is $4/52$, which is our P(A). Given that the first card is an ace, there are now three aces left among the remaining 51 cards. The conditional probability of another ace is thus $3/51$, our P(B given A). Multiply the two to get the probability of two aces as $4/52 \times 3/51$, which is approximately 0.0045.

If you compare the two versions of the multiplication rule, you realize that independent events have the special property that P(B) = P(B given A); the unconditional and the conditional probabilities are the same. This observation makes sense. In the last example, suppose that instead of drawing two cards, you draw one card, put it back, reshuffle the deck, and draw again. Now what is the probability that you get two aces? In this case, the events to get an ace in the first and second draw *are* independent and the probability is $4/52 \times 4/52$, which is about 0.0059. (Why is it larger than before? Think about what happens if you draw three, four, or five cards and ask for three, four, or five aces.)

Note that there is a certain symmetry here. If P(B given A) is different from P(B), then P(A given B) is also different from P(A). You may try to prove this rule from the multiplication formula above, noting that "A and B" is the same as "B and A." Also note that the multiplication rule gives a way to compute the conditional probability if it is not obvious how to do so directly. Shuffling around the factors in the formula above gives the expression

$$\text{P(B given A)} = \frac{\text{P(A and B)}}{\text{P(A)}}$$

which will be useful to us later. Note the difference between P(A and B), which is the probability that *both* A and B occur, and P(B given A), which is the probability that B occurs *if we know that* A *has occurred.* These can be quite different. For example, choose an American at random. Let A be the event that you get somebody from Rhode Island and B the event that you get somebody of Portuguese descent (a *Luso-American*). Then P(A and B) is the probability to get a Rhode Islander with Portuguese ancestry, which is about 0.03% (there are about 90,000 such individuals among the U.S. population of 295 million). The conditional probability P(B given A), on the other hand, is the probability that a Rhode Islander has Portuguese ancestry, and this is about 9% (90,000 out of a million).

For a simple illustration of how to compute conditional probabilities, let us again turn to the experiment of rolling two dice. Let A be the event to get at least one 6 and let B be the event that the sum of the dice equals ten. See Figure 1.5 for the sample space with these two events. Two outcomes satisfy both A and B ("4 on first, 6 on second" and "6 on first, 4 on second"), and we thus have P(A and B) $= 2/36$. In the figure you also see that P(A) $= 11/36$, and by the formula above, the conditional probability that the sum is ten given that at least one die shows 6 is

$$\text{P(B given A)} = \frac{\text{P(A and B)}}{\text{P(A)}} = \frac{2/36}{11/36} = 2/11$$

and you can understand this intuitively: If you know that there is at least one 6, there are 11 possible outcomes, and because two of these have the sum equal to ten, the conditional probability to get sum ten is $2/11$. It is nice to see that the formal computation and the intuitive reasoning agree. Provided that your intuition does not go agley, they always do.

In the early 1990s, a leading Swedish tabloid tried to create an uproar with the headline "Your ticket is thrown away!". This was in reference to the popular

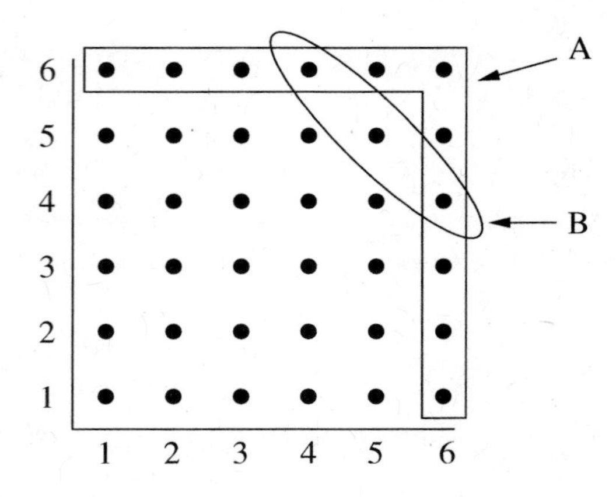

Figure 1.5 Rolling two dice. The events marked are A: to get at least one 6, and B: to get a sum equal to ten.

Swedish TV show "Bingolotto" where people bought lottery tickets and mailed them to the show. The host then, in live broadcast, drew one ticket from a large mailbag and announced a winner. Some observant reporter noticed that the bag contained only a small fraction of the hundreds of thousands tickets that were mailed. Thus the conclusion: Your ticket has most likely been thrown away!

Let us solve this quickly. Just to have some numbers, let us say that there are a total of 100,000 tickets and that 1,000 of them are chosen at random to be in the final drawing. If the drawing was from all tickets, your chance to win would be 1/100,000. The way it is actually done, you need to both survive the first drawing to get your ticket into the bag and then get your ticket drawn from the bag. The probability to get your entry into the bag is 1,000/100,000. The conditional probability to be drawn from the bag, given that your entry is in it, is 1/1,000. Multiply to get 1/100,000 once more. There were no riots in the streets.

Conditional probability can also explain why Mrs. T from page 11 made her statement "That sounds more likely." She thought of a conditional probability without even knowing it. It was hard for her to believe that a former senator could be so shabby, but when she found out more about him, she found it easier to believe. Thus, in her mind, P(B given A) was larger than P(B) (what are A and B?).

Misunderstanding probability can be more serious than upsetting Swedish TV viewers or making fun of Louisiana politicians. Just ask Sally Clark. In

1999, a British jury convicted her of murdering two of her children who had died suddenly at the ages of 11 and 8 weeks, respectively. A pediatrician called in as an expert witness claimed that the chance of having two cases of infant sudden death syndrome, or "cot deaths," in the same family was 1 in 73 million. There was no physical or other evidence of murder, nor was there a motive. Most likely, the jury was so impressed with the seemingly astronomical odds against the incidents that they convicted. But where did the number come from? Data suggested that a baby born into a family similar to the Clarks faced a 1 in 8,500 chance of dying a cot death. Two cot deaths in the same family, it was argued, therefore had a probability of $1/8{,}500 \times 1/8{,}500$, which is roughly equal to $1/73{,}000{,}000$.

Did you spot the error? I hope you did. The computation assumes that successive cot deaths in the same family are independent events. This assumption is clearly questionable, and even a person without any medical expertise might suspect that genetic factors play a role. Indeed, it has been estimated that if there is one cot death, the next child faces a much larger risk, perhaps around $1/100$. To find the probability of having two cot deaths in the same family, we should thus use conditional probabilities and arrive at the computation $1/8{,}500 \times 1/100$, which equals $1/850{,}000$. Now, this is still a small number and might not have made the jurors judge differently. But what does the probability $1/850{,}000$ have to do with Sally's guilt? Nothing! When her first child died, it was certified to have been from natural causes and there was no suspicion of foul play. The probability that it would happen again without foul play was $1/100$, and if that number had been presented to the jury, Sally would not have had to spend three years in jail before the verdict was finally overturned and the expert witness (certainly no expert in probability) found guilty of "serious professional misconduct."

You may still ask the question what the probability $1/100$ has to do with Sally's guilt. Is this the probability that she is innocent? Not at all. That would mean that 99% of all mothers who experience two cot deaths are murderers! The number $1/100$ is simply the probability of a second cot death, which only means that among all families who experience one cot death, about 1% will suffer through another. If probability arguments are used in court cases, it is very important that all involved parties understand some basic probability. In Sally's case, nobody did.

Next, let us look at a paradox that is not usually presented as a probability problem. Your teacher tells the class there will be a surprise exam next week. On one day, Monday–Friday, you will be told in the morning that an exam is to be given on that day. You quickly realize that the exam will not be given on Friday; if it was, it would not be a surprise because it is the last possible day to get the exam. Thus, Friday is ruled out, which leaves Monday–Thursday. But then Thursday is impossible also, now having become the last possible day to get the exam. Thursday is ruled out, but then Wednesday becomes impossible, then Tuesday, then Monday, and you conclude: There is no such thing as a surprise exam! But the teacher decides to give the exam on Tuesday, and come Tuesday morning, you are surprised indeed.

This problem, which is often also formulated in terms of surprise fire drills or surprise executions, is known by many names, for example, the "hangman's paradox" or by serious philosophers as the "prediction paradox." To resolve it, I find it helpful to treat it as a probability problem. Let us suppose that the day of the exam is chosen randomly among the five days of the week. Now start a new school week. What is the probability that you get the test on Monday? Obviously $1/5$ because this is the probability that Monday is chosen. If the test was not given on Monday, what is the probability that it is given on Tuesday? The probability that Tuesday is chosen to start with is $1/5$, but we are now asking for the *conditional* probability that the test is given on Tuesday, given that it was not given on Monday. As there are now four days left, this conditional probability is $1/4$. Similarly, the conditional probabilities that the test is given on Wednesday, Thursday, and Friday conditioned on that it has not been given thus far are $1/3$, $1/2$, and 1, respectively.

We could define the "surprise index" each day as the probability that the test is *not* given. On Monday, the surprise index is therefore 0.8, on Tuesday it has gone down to 0.75, and it continues to go down as the week proceeds with no test given. On Friday, the surprise index is 0, indicating absolute certainty that the test will be given that day. Thus, it is possible to give a surprise test but not in a way so that you are equally surprised each day, and it is never possible to give it so that you are surprised on Friday.

LIAR, LIAR

This entire section is devoted to a classic probability problem. It is easy to state but can lead to great confusion and frustration. It may strike you as a wee bit tedious, and if you feel that it fails to catch your interest, you can safely skip this section and proceed to the next without missing any vital information.

The problem is a typical example of how you sometimes need to stop and think about what you are asked to do before you do anything. It goes like this: Adam, Bob, and Carol are each known to tell the truth with probability $1/3$ (independently of each other) and lie otherwise. If Adam denies that Bob confirms that Carol lies, what is the probability that Carol tells the truth?

Yikes. Let us first realize that we are here in fact asked for a conditional probability and name the two events of interest:

C: Carol tells the truth
A: Adam denies that Bob confirms that Carol lies

We are now asking for the conditional probability P(C given A). From page 21, we know that this can be computed as P(C and A) divided by P(A), so let us first find P(C and A). Let us get rid of a double negation and rephrase A as

Adam says that Bob says that Carol tells the truth

which means that the combined event "C and A" can be written as

Carol tells the truth and Adam says that Bob says that Carol tells the truth

The question is now: For which combinations of lying and truth-telling among the three will this last event occur? First of all, Carol must tell the truth. What about the others? If Adam tells the truth when he confirms that Bob confirms Carol's truth-telling, then Bob is also telling the truth. Thus, the combined event occurs if everybody tells the truth, and this has probability $1/3 \times 1/3 \times 1/3 = 1/27$.

What if Adam lies? Then Bob says that Carol lies, so he is also lying and the combination "Adam lies, Bob lies, Carol tells the truth" also makes the combined event occur. That combination has probability $2/3 \times 2/3 \times 1/3 = 4/27$. Adding this number to the $1/27$ from above and noting that no other combinations work, we conclude that

$$P(\text{C and A}) = 5/27$$

For the event A alone, the two combinations above make it occur but there are more possibilities. For example, if Adam tells the truth and both Bob and Carol lies, the event A occurs. Why? Well, suppose that Adam truthfully confirms that Bob says: "Carol tells the truth." If Bob lies, this means that Carol also lies and the event "Adam says that Bob says that Carol tells the truth" occurs. Table 1.1 gives all possible combinations of truth-telling (T) and lying (L), and whether the events occur. Note that for the event A to occur, we need an odd number of truth-tellers, and for the combined event "C and A" to occur, we in addition need Carol to be one of them.

Table 1.1 Possible combinations of lying and truth-telling for the three individuals

Adam	T	T	T	T	L	L	L	L
Bob	T	T	L	L	T	T	L	L
Carol	T	L	T	L	T	L	T	L
A	yes	no	no	yes	no	yes	yes	no
C and A	yes	no	no	no	no	no	yes	no
Probability	$1/27$	$2/27$	$2/27$	$4/27$	$2/27$	$4/27$	$4/27$	$8/27$

We computed P(C and A) above, and by adding the probabilities of the "yes" entries in the table, we also get $P(A) = 13/27$. We can now compute the desired conditional probability as

$$P(\text{C given A}) = \frac{P(\text{C and A})}{P(\text{A})} = \frac{5/27}{13/27} = 5/13$$

Note that this probability is about 38.5%, slightly higher than the 33.3% that is the unconditional probability that Carol tells the truth. The fact that Adam says that Bob confirms Carol's truth-telling makes us believe in her a little more, which might be a bit surprising because these guys are such a bunch of liars.

This problem is an old one. It was published by British astrophysicist Sir Arthur Eddington in his 1935 book *New Pathways in Science* and further explained by him in a 1935 article in the *The Mathematical Gazette*. He claims in turn to have learned about it in a 1919 after-dinner speech by his colleague A. C. D. Crommelin (who has a comet named after him). In Sir Arthur's version, which I will state shortly, there is a fourth person also involved. Interestingly, the problem led to some controversy and different solutions were published. This discrepancy has to do with one crucial assumption that I made above but did not explicitly state: If Adam lies when he says "Bob says that Carol tells the truth," I interpreted this as meaning that Bob says that Carol lies, but it could also mean that Bob did not say anything at all. In fact, the whole solution rests on the following chain of assumptions that are not spelled out in the problem: First Carol says something that is either true or false. Next, Bob who knows whether she told the truth, says either "Carol tells the truth" or "Carol lies." Finally, Adam says either "Bob says that Carol tells the truth" or "Bob says that Carol lies." This interpretation also validates my rewriting to "get rid of a double negation."

However, if we return to the original formulation "Adam denies that Bob confirms that Carol lies," it might also be argued that Adam is asked the question, "Does Bob confirm that Carol lies?" and answers, "No." If Adam lies, it means that Bob does indeed say, "Carol lies." However, if Adam speaks the truth, this could mean that Bob denies that Carol lies, but it could also mean that Bob has not said anything at all in the matter. The latter was Sir Arthur's interpretation of the problem. His only assumption was that Carol makes a statement that is either true or false, which led him to exclude only cases that are clearly inconsistent with the statement in the problem formulation (his interest in the problem in the first place was as an illustration of what he called the "exclusion method" in interpreting observational results in physics). In his view, the only cases inconsistent with the statement A are the cases L–T–T and L–L–L in the notation of the table above. The cases T–T–L and T–L–T, which we excluded from A are included by Sir Arthur. He views all cases in which Adam tells the truth as consistent with A; if Adam tells the truth, Sir Arthur argues, we simply cannot say anything about what Bob has said and there is no evidence against Carol speaking the truth. In his interpretation, $P(C \text{ and } A) = 7/27$ and $P(A) = 17/27$, which gives the final answer that Carol tells the truth with probability $7/17$.

In the December 1936 issue of the *Gazette*, two articles were published: one that agreed with Sir Arthur and one that disagreed. Of course there is no universally correct answer, only a correct answer relative to the assumptions that are made. With Sir Arthur's interpretation, one must assume that Carol tells the truth when it cannot be proved that she lies. The interpretation depends also on what context we imagine. If these people are testifying in a court trial, it is reasonable to assume that they have all made statements and our interpretation is logical. If it is instead an illustration of some principle in physics, Sir Arthur's interpretation may perhaps make sense, but I believe that most probabilists would agree that the only way to properly solve the problem is to make the assumptions that we have made. Still, I would not go as far as Warren Weaver, who in his modern classic *Lady Luck: The Theory of Probability* claims that Sir Arthur's solution involves a condition that is "rather ridiculous in character" (*Lady Luck* that came out in 1963 is by the way a wonderful book, to this day arguably the best non-technical introduction to probability that there is). Of course, the complete set of assumptions in the liar problem should not be spelled out in the problem formulation. That would clutter the problem and reduce the chance of interesting conflicts between feisty British astrophysicists. Here is Sir Arthur's original formulation:

> If A, B, C, and D each speak the truth once in three times (independently), and A affirms that B denies that C declares that D is a liar, what is the probability that D was speaking the truth?

I will leave it as an exercise for you to solve it. Sir Arthur gave the answer $25/71$. With our interpretation, the correct answer is $13/41$.

TOTAL PROBABILITY, USED CARS, AND TENNIS MATCHES

Suppose that you buy a used car in a city where street flooding is a common problem. You know that roughly 5% of all used cars have been flood-damaged and estimate that 80% of such cars will later develop serious engine problems, whereas only 10% of used cars that are not flood-damaged develop the same problems. Of course, no used car dealer worth his salt would let you know whether your car has been flood damaged, so you must resort to probability calculations. What is the probability that your car will later run into trouble?

You might think about this problem in terms of proportions. Out of every 1,000 cars sold, 50 are previously flood-damaged, and of those, 80%, or 40 cars, develop problems. Among the 950 that are not flood-damaged, we expect 10%, or 95 cars, to develop the same problems. Hence, we get a total of $40+95 = 135$ cars out of a thousand, and the probability of future problems is 13.5%.

If you solved the problem in this way, congratulations. You have just used the *law of total probability*, one of the most useful rules that we have in probability. If we restate everything in terms of probabilities, we have done the calculation

$$\text{P(engine problems)} = 0.05 \times 0.80 + 0.95 \times 0.10 = 0.135$$

which means that we considered the two different cases "flood damaged" and "not flood damaged" separately and then combined the two to get our probability. For those of you who like math formulas, here is a general formula for two events events A and B:

$$P(B) = P(B \text{ given } A) \times P(A) + P(B \text{ given (not } A)) \times P(\text{not } A)$$

It is Sunday evening down at the local pub. You and your two colleagues Albert and Betsy have met for a pint and start discussing the bus you catch to work every morning, which tends to be late about 40% of the time. You decide to see who can best predict whether the bus will be late or on time for the entire next week. Each of you will suggest a sequence of five Ls and Ts (for "Late" and "on Time"). As you know that the bus is late with probability 0.4 each day, you decide to generate a sequence at random by choosing L with probability 0.4 and T with probability 0.6, five times. Your friend Albert thinks along the same lines but does not want to run the risk of getting too many Ls so he decides on a sequence with two Ls and three Ts and chooses their positions randomly. Betsy notes that each day the bus is more likely than not to be on time and simply suggests the sequence TTTTT (which prompts Albert to shake his head and sigh "women" in his pint, "of course it will not be on time every day"). Who is most likely to guess the entire week correctly?

Let us compute the probability to guess correctly for a single day. If your guess is T, you are correct if the bus is on time, which has probability 0.6. Thus, Betsy has this probability every day, and Albert has it three days and 0.4 the other two days. You, with your more complicated strategy, are correct if you guess T and the bus is on time or if you guess L and the bus is late. The law of total probability gives that you are correct with probability $0.6 \times 0.6 + 0.4 \times 0.4 = 0.52$. With independence between the five days of the week, you have probability $0.52^5 \approx 0.038$ to be correct. Albert's probability is $0.6^3 \times 0.4^2 \approx 0.035$ (this number is the same regardless of which two days he chooses for his two Ls) and Betsy's probability is $0.6^5 \approx 0.078$. With probabilities 3.5%, 3.8%, and 7.8%, neither of you has a very good chance to get it right, but Betsy definitely has an edge. To somewhat save his face, Albert can claim that he is more likely than Betsy to get the *number* of late days right. The probability that there are two late days is about 0.35 (which is his 0.035 multiplied by 10; can you see why the relevant probability is computed in this way?). His problem is, of course, to find the right two days.

The law of total probability also works with more than two events. Suppose that Ann and Bob play a game of tennis and that Ann is about to serve at deuce, which means that whoever first gets two points ahead wins the game. Suppose that Ann wins a point with probability $2/3$. What is the probability that she wins the game?

Not so easy to figure out directly. In fact, there is an unlimited number of ways in which Ann can win. She can win two straight points. She can win a point, lose a point, and then win two straight points. She can win, lose, win,..., lose, and then win two straight points. Probabilists do not fear infinite sums, and it is possible to find the probability that Ann wins the game by computing and adding the probabilities of all these cases. Feel free to try it for yourself. I will, however, demonstrate a more elegant way. Let us consider three distinct cases:

Case I: Ann wins the next two points

Case II: Bob wins the next two points

Case III: They win a point each, in any order

Now use the law of total probability with the three cases I, II, and III to get the formula

$$\begin{aligned}\text{P(Ann wins)} &= \text{P(Ann wins in I)} \times \text{P(I)} + \text{P(Ann wins in II)} \times \text{P(II)} \\ &\quad + \text{P(Ann wins in III)} \times \text{P(III)}\end{aligned}$$

What are now all these probabilities? We assume independence between consecutive points, which gives that the cases have probabilities

$$\text{P(I)} = 2/3 \times 2/3 = 4/9$$

$$\text{P(II)} = 1/3 \times 1/3 = 1/9$$

$$\text{P(III)} = 4/9$$

where the last probability is computed by adding the first two and subtracting the sum from one. So far so good. Now for the conditional probabilities. The first two are easy; clearly P(Ann wins in I) $= 1$ and P(Ann wins in II) $= 0$. But what about the third? In case III, the players are back at deuce, so the probability that Ann wins is now precisely the probability we asked for in the first place. Are we stuck?

No, in fact we are almost done! As P(Ann wins) and P(Ann wins in III) are equal but unknown, let us call this unknown number p and plug it in above together with the known probabilities:

$$
\begin{aligned}
p &= 1 \times 4/9 + 0 \times 1/9 + p \times 4/9 \\
&= 4/9 + p \times 4/9
\end{aligned}
$$

This formula is now an equation for p that is easy to solve and has the solution $p = 4/5$. The probability that Ann wins is $4/5$. Instead of computing the probability directly, we found three different cases where two could be dealt with explicitly and the third brought us back to the beginning, giving a simple equation for the unknown probability. Wasn't that cool? A general formula if Ann wins a point with probability w is given by

$$
\text{P(Ann wins)} = \frac{w^2}{w^2 + (1-w)^2}
$$

which you can try to deduce on your own using the same idea as above. See Figure 1.6 for an illustration of the different cases in a *tree diagram*.

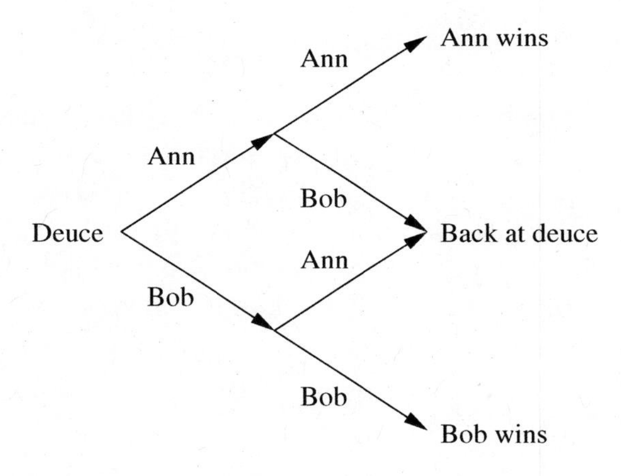

Figure 1.6 The different scenarios when Ann and Bob start from deuce.

Let us finish by another racket sport problem, this time regarding badminton. In the United States this sport is mostly considered a backyard game, and if you go by the badminton courts in a college gym, about 90% of the players are Asian, the rest being Scandinavian, with the odd Brit, German, or New Zealander tossed in the mix. However, in August 2005, Howard Bach and Tony Gunawan made history by winning the men's doubles gold medal in the world championships, the first U.S. players ever to become world champions. It may be surprising to many that badminton is actually the fastest racket sport.

The shuttlecock can reach top speeds of 200 mph, which is not bad for a bunch of goose feathers stuck into a cork.

In badminton you can only score when you serve. The exchange of shots is called a "rally"; thus, if you win a rally as server, you score a point. If you win a rally as receiver, the score is unchanged but you get to serve and the opportunity to score. Suppose that Ann and Bob are equally strong players so that Ann wins a rally against Bob with probability $1/2$ regardless of who serves (a reasonable assumption in badminton, but would of course not be so in tennis where the server has a big advantage). What is the probability that Ann scores the next point if she is the server?

We will use the same idea as in the tennis example. This time, the three cases are as follow as follows:

Case I: Ann wins the rally

Case II: Ann loses the next two rallies

Case III: Ann loses the rally and wins the following rally.

In case I, Ann scores the point; in case II, Bob scores the point; and in case III, they are back where they started with Ann serving again, no point scored yet. The cases have probabilities $\mathrm{P(I)} = 1/2, \mathrm{P(II)} = 1/4$, and $\mathrm{P(III)} = 1/4$, and the first two conditional probabilities are P(Ann wins in I) $= 1$ and P(Ann wins in II) $= 0$. The third conditional probability, P(Ann wins in III), is equal to the original probability P(Ann wins), so we denote this unknown number by p and get the equation

$$\begin{aligned} p &= 1 \times 1/2 + 0 \times 1/4 + p \times 1/4 \\ &= 1/2 + p \times 1/4 \end{aligned}$$

which has the solution $p = 2/3$. Thus, the server has a notable advantage and this is even more pronounced if one player is a little better than the other. For example, if Ann has probability 0.55 to win a rally, her probability to win the next point if she is currently the server goes up to 0.73. This phenomenon may explain why seemingly extreme set scores such as 15–3 or 15–4 are not uncommon in badminton tournaments. The general formula this time is given by

$$\mathrm{P(Ann\ wins)} = \frac{w}{w + (1 - w)^2}$$

where w is the probability that Ann wins a rally. Again, I leave it as an exercise for you to deduce the formula.

COMBINATORICS, PASTRAMI, AND POETRY

Combinatorics is the mathematics of counting and something that shows up in many probability problems. The fundamental principle in combinatorics is the *multiplication principle*, which is easier to illustrate with examples than try to state formally. Let's do lunch. Suppose that a deli offers three kinds of bread, three kinds of cheese, four kinds of meat, and two kinds of mustard. How many different meat and cheese sandwiches can you make? First choose the bread. For each choice of bread, you then have three choices of cheese, which gives a total of $3 \times 3 = 9$ bread/cheese combinations (rye/swiss, rye/provolone, rye/cheddar, wheat/swiss, wheat/provolone,...you get the idea). Then choose among the four kinds of meat, and finally between the two types of mustard or no mustard at all. You get a total of $3 \times 3 \times 4 \times 3 = 108$ different sandwiches. Suppose that you also have the choice of adding lettuce, tomato, or onion in any combination you want. This choice gives another $2 \times 2 \times 2 = 8$ combinations (you have the choice "yes" or "no" three times) to combine with the previous 108, so the total is now $108 \times 8 = 864$.

That was the multiplication principle. In each step you have several choices, and to get the total number of combinations, multiply. It is fascinating how quickly the number of combinations grow. Just add one more type of bread, cheese, and meat, respectively, and the number of sandwiches becomes 1,920. It would take years to try them all for lunch.

Another example is to consider how many possible positions there are in chess after two moves. White starts and has 20 possible opening moves. For each of these, black also has 20 possible moves and there are thus $20 \times 20 = 400$ possible positions already after the first two moves (but only a few of these would ever show up in a serious game). After the two opening moves, the number of possible moves depends on the previous moves, but suffice it to say that the number of positions grows very rapidly. No wonder computers are better chess players than people (sorry chess players). A somewhat related example that I am sure you have heard is the tale of the king who agreed to award the inventor of chess by placing one grain of rice on the first square, two on the second, and keep on doubling until the board was full. The last square would then have $2 \times 2 \times \cdots \times 2 = 2^{63}$ grains of rice, which would make enough sushi to feed the entire world for many years.

To provide a link between probability and poetry, we turn to the French poet and novelist Raymond Queneau who in 1961 wrote a book called *One Hundred Thousand Billion Poems*. The book has ten pages, and each page contains a sonnet, which has 14 lines. There are cuts between the lines so that each line can be turned separately, and because all lines have the same rhyme scheme and rhyme sounds, any such combination gives a readable sonnet. The number of sonnets that can be obtained in this way is thus 10^{14}, which is indeed a hundred thousand billion. Somebody has calculated that it would take about 200 million years of nonstop reading to get through them all. I would instead recommend Queneau's hilarious *Exercises in Style* in which the same story is retold in 99 different styles and which can be read in an afternoon. One may wonder if Queneau was ever asked what he thought was his best work and replied "I don't know. I haven't read most of it."

How does probability enter into all this? Here is an example. A Swedish license plate consists of three letters followed by three digits. What is the probability that a randomly chosen such plate has no duplicate letters and no duplicate digits?

The first question is how many letters there are in the Swedish alphabet. Aren't there letters like å, ä, and ö? Yes, but these are not used for license plates. A few others are also not used, and the total number of available letters is 23. There is, therefore, a total of $23 \times 23 \times 23 \times 10 \times 10 \times 10$, which is approximately 12 million license plates (excluding typical Swedish vanity plates such as VIKING or I♡ABBA). To get a plate that has no duplicate letters, we can choose the first letter in any way we want, so for this we have 23 choices. The next letter cannot be the same as the first, so here we have 22 choices. Finally, the third letter cannot be equal to any of the first two, which gives 21 choices. Same for the digits: first 10, then 9, and then 8 choices. The number of plates with no duplicates is thus $23 \times 22 \times 21 \times 10 \times 9 \times 8$. Divide by the total number to get the probability

$$P(\text{no duplicate letters or digits}) = \frac{23 \times 22 \times 21 \times 10 \times 9 \times 8}{23 \times 23 \times 23 \times 10 \times 10 \times 10} \approx 0.63$$

For the license plates, the order is important. For example, ABC123 is different from BCA231. In some combinatorial problems, order is irrelevant, for example, those that have to do with poker. You are dealt a poker hand (5 cards from a regular deck of 52 cards). What is the probability of being dealt a *flush* (five cards in the same suit, not all five consecutive)?

First find the total number of different hands, and then the number of hands that give a flush. As there are 52 ways to choose the first card, 51 ways to choose the second, and so on down to 48 ways to choose the fifth, the multiplication principle tells us that there are $52 \times 51 \times \cdots \times 48 = 311{,}875{,}200$ different ways to get your cards. But then we have taken order into account and for example distinguished between the sequences (♠A, ♢A, ♠2, ♡A, ♣A) and (♣A, ♠2, ♢A, ♠A, ♡A), and in poker, four aces are four aces regardless of order. As there are $5 \times 4 \times 3 \times 2 \times 1 = 120$ ways to rearrange any given five cards (five choices for the first, four for the second, and so on), we need to divide 311,875,200 by 120 to get the number 2,598,960. There are about 2.6 million different poker hands.

For the number of hands that give a flush, first consider a flush in some given suit, for example, hearts. For the first card, we then have 13 choices, for the second 12, and so on, and each such sequence of five cards can be rearranged in 120 ways, which gives $13 \times 12 \times 11 \times 10 \times 9/120 = 1{,}287$ hands that give five hearts. From this number we must subtract the ten hands that have five consecutive hearts, because such a hand counts as a *straight flush*, not a mere flush. The subtraction leaves 1,277 hands with a flush in hearts. Finally, as there are four suits, the number of hands that give a flush in any suit is $4 \times 1{,}277 = 5{,}108$. The probability to be dealt a flush is therefore $5{,}108/2{,}598{,}960 \approx 0.002$. You are dealt a flush on average once every 500 hands .

Let us introduce some notation that is convenient to use. You may be familiar with the notation for the *factorial* of a number n:

$$n! = n \times (n-1) \times \cdots \times 2 \times 1$$

Thus, $1! = 1, 2! = 2, 3! = 6, 4! = 24, 5! = 120$, and $6! = 720$. The exclamation mark is not intended to indicate surprise, but factorials do grow surprisingly quickly. For example, the total number of ways of rearranging a deck of cards is 52!, which is an enormous number. Take out a deck and shuffle it. Do you think that the particular order of the cards that you got has ever occurred before in a deck in the history of card playing? Most likely not. If all 6.5 billion people on earth started shuffling cards and produced one shuffled deck every ten seconds around the clock they would have to do this for about four million sextillion septillion years to even have a chance of producing all possible orders. That's a number with 51 digits. That's a long time.

We saw above that the number of possible poker hands, that is, the number of ways to choose five cards out of 52 is $52 \times 51 \times \cdots \times 48/5!$. In general, if we choose k out of n objects, there are $n \times (n-1) \times \cdots \times (n-k+1)/k!$ ways to do this (convince yourself!). We use the following special notation:

$$\binom{n}{k} = \frac{n \times (n-1) \times \cdots \times (n-k+1)}{k!}$$

a number that is read "n choose k." If the numerator looks messy to you, just remember that it has k factors. Thus, there are $\binom{52}{5}$ different poker hands, $\binom{13}{5} - 10$ hands that give a flush in hearts, and $4 \times (\binom{13}{5} - 10)$ hands that give you a flush in any suit. Check that this agrees with our calculations above. Also convince yourself of the following identity:

$$\binom{n}{k} = \binom{n}{n-k}$$

which can come in handy in computations. For example, if you have to compute $\binom{10}{8}$ by hand, it is easier to instead compute $\binom{10}{2}$ (do it and you will see why). A quick argument for the formula is that each choice of k objects can also be done by setting aside the $n-k$ objects that you do *not* choose. As there are $\binom{n}{k}$ ways to do the first and $\binom{n}{n-k}$ ways to do the second, the two expressions must be the same. Needless to say, computations by hand are seldom done these days, and even fairly simple pocket calculators have functions to compute $\binom{n}{k}$. The formula is still good to know.

Here is a real-life problem that comes from the field of home care medicine. It has been observed that the risk of a drug interaction is about 6% for a patient who takes two medications and about 50% for a patient who takes five medications. What is the risk of a drug interaction for a patient who takes nine medications? First of all, with nine medications there are $\binom{9}{2} = 9 \times 8/2 = 36$ pairs of medications that can interact. We are now looking for the probability that at least one such pair leads to an interaction and Trick Number One comes in handy again. The probability that there is no interaction between any two medications is 0.94, and if we assume that pairs are independent of each other, the probability of no interaction is

$$\text{P(no interaction)} = 0.94^{36} \approx 0.11$$

and the risk of having an interaction is thus $1 - 0.11 = 0.89$, almost 90%. One can question whether the independence assumption is reasonable; if medication A interacts with medication B, perhaps it is also more likely to interact with other medications. To test this assumption, we can use the other piece of information given, that the risk is about 50% for those who take five medications. With five medications, there are $\binom{5}{2} = 5 \times 4/2 = 10$ pairs, and the risk of interaction is $1 - 0.94^{10} \approx 0.46$ under the independence assumption. The 46% is close enough to the observed 50% to motivate our assumption. This problem was kindly given to me by one anonymous reviewer of my book proposal. Thanks reviewer number three.

A more playful problem now. Roll a die six times. What is the probability that all six sides come up? At first glance, this problem does not seem to have anything to do with combinatorics, but we can translate it into a situation with balls and urns. Let us thus consider six urns, labeled 1 through 6 and six balls, numbered 1 through 6. Roll the die. If it shows 1 put ball number 1 in urn number 1; if it shows 2, put ball number 1 in urn number 2, and so on. Roll the die again, and do the same thing with ball number 2. After all six rolls, the six balls are distributed among the urns. The total number of ways in which this can be done is 6^6 by the multiplication principle (six choices of urn for each of the six balls). To get all different numbers, there are 6 choices for the first ball, 5 for the second, and so on; thus, a total of $6!$ ways. We have argued that the probability that all six sides come up is

$$\text{P(all six sides)} = 6!/6^6 \approx 0.015$$

A general formula for the probability that n balls distributed over n urns leaves no urn empty is thus

$$\text{P(no urn empty)} = n!/n^n$$

and even though the factorial in the numerator grows fast, it stands no chance against the denominator. The probability rapidly approaches 0 as n increases.

THE VON TRAPPS AND THE BINOMIAL DISTRIBUTION

Recall the problem on page 6 where we asked for the probability that a family with three children has exactly one daughter. By listing the eight possible outcomes and counting the cases with one daughter, we arrived at the solution $3/8$. Another way to solve this problem is to first note that by independence, each particular sequence of two boys and one girl, for example BBG, has probability $1/2 \times 1/2 \times 1/2 = 1/8$. As there are three such sequences (GBB, BGB, and BBG), we get the probability $3 \times 1/8 = 3/8$.

Let us now instead consider a family with seven children. What is the probability that they, like the von Trapps, have five daughters? There are now $2^7 = 128$ possible outcomes ranging from BBBBBBB to GGGGGGG. It is tedious to list them all and count how many that have five girls. Let us instead try the second approach. Each particular sequence with five girls and two boys, for example, GBGGBGG, has probability $(1/2)^7$ so the question is how many such sequences there are. Here is where combinatorics come in. The question becomes: In how many ways can we choose positions for the five Gs? The answer is $\binom{7}{5}$. Recall from the previous section that it is easier to compute this as $\binom{7}{2}$, which equals $7 \times 6/2 = 21$. The probability that a family with seven children has five daughters is thus

$$P(\text{five daughters}) = 21 \times (1/2)^7 \approx 0.16$$

We've done the von Trapps. Now you do the Jacksons. What is the probability that a family with nine children has three daughters?[5]

Here is another problem that can be solved in the same way. If you roll a die 12 times, you expect to get on average two 6s but what is the probability to get exactly two 6s? First think of a particular sequence of 12 rolls with two 6s, for example, XX6X6XXXXXXX, where "X" means "something else." By independence we multiply and get the probability $5/6 \times 5/6 \times 1/6 \times \cdots \times 5/6$, which we can also write as $(1/6)^2 \times (5/6)^{10}$. But this is the probability for any specified sequence with two 6s, so the question again is: How many sequences are there? And just like above, we need to choose positions, this time for two 6s in a sequence of 12 rolls. We get

$$P(\text{two 6s in twelve rolls}) = \binom{12}{2} \times (1/6)^2 \times (5/6)^{10} \approx 0.3$$

If we were to use the "old method" of counting in the sample space, we must first note that because 6 and X are not equally likely, we cannot use the sample space of the $2^{12} = 4{,}096$ outcomes ranging from XXXXXXXXXXXX to 666666666666. We have to break up each X into five different outcomes, combine all of these, and end up with a sample space with 6^{12}, a bit over

[5]For bonus points, name the daughter that is not Janet or La Toya.

two billion, equally likely outcomes. It is possible to proceed and solve the problem in this way, but I would not recommend it as a general method.

Let us do the general formula now. Suppose that an experiment (such as giving birth or rolling a die) is repeated n times. We refer to each repetition of the experiment as a *trial*. Each trial results in a "success" with some probability p, independently of previous trials. The probability to get exactly k successes is then

$$\mathrm{P}(k \text{ successes}) = \binom{n}{k} \times p^k \times (1-p)^{n-k}$$

where k can be anything from 0 to n. So that the formula makes sense for $k = 0$ and $k = n$, the number $\binom{n}{0}$ is defined to be equal to 1, and any number raised to the power 0 is also defined to be 1. The probability to get 0 successes is then $(1-p)^n$, and the probability to get all n successes is p^n. We say that the number of successes has a *binomial distribution* (the numbers $\binom{n}{k}$ are called *binomial coefficients* and you may be familiar with *Newton's binomial theorem*). The numbers n and p are called the *parameters* of the binomial distribution. In the von Trapp example, we have $n = 7$ and $p = 1/2$; in the dice example, $n = 12$ and $p = 1/6$. In the problem with drug interactions on page 36, the number of interactions with nine medications has a binomial distribution with $n = 36$ and $p = 0.06$, and we computed the probability for $k = 0$.

Two assumptions are crucial for the binomial distribution. First, successive trials must be independent of each other, and second, the success probability p must be the same in each trial. Let me illustrate this in an example. Call a day "hot" if the high temperature is above 90 degrees, and suppose that the probability of a hot day in New Orleans in early July is 0.7. You now decide to count the number of hot days among

(a) Each Fourth of July for the next five years

(b) Each day in the first week of July next year

(c) Each first day of the month next year

Does any of (a), (b), or (c) give you a binomial distribution for the number of hot days?

The answer is that only (a) gives a binomial distribution. The temperature on July 4th one year is certainly independent of the temperature on July 4th another year, and it is reasonable to assume that our success probability 0.7 stays the same for another five years (global warming isn't *that* fast). Thus, you have a binomial distribution with parameters $n = 5$ and $p = 0.7$. In (b), the trials are not independent. If you have a hot day on July 1, you are more likely to get a hot day also on July 2 because there is a weather system in place that gives you hot temperatures. Thus, you do not have a binomial distribution in this case. The fact that an arbitrary day in early July is hot with probability 0.7 means that over the years, about 70% of early July days have been hot. These hot days have typically been concentrated to certain years though. Some years all or most days in early July were hot, and some years there were few, if any. Thus, on average, seven out of ten days are hot, but consecutive days are not independent.

In (c) finally, although it might be reasonable to assume that the temperature on the first of one month is independent of the temperature on the first of another month (weather systems don't usually stay around for that long), the problem is that the success probability changes. The probability of a hot day is, for example, far less than 0.7 in January, so you do not have a binomial distribution here either.

Recall the pub evening from page 29, but let us suppose that you instead use your strategies to guess the number of days the bus is late. The number of such late days has a binomial distribution with the parameters $n = 5$ (the number of work days in a week) and $p = 0.4$ (the probability that the bus is late on any given day). This time Betsy fares the worst. The probability that she is correct is the probability that the bus is never late, which corresponds to $k = 0$ in the formula and gives probability $0.6^5 \approx 0.08$. Albert is correct if the bus is late twice and the probability of this is

$$\text{P(the bus is late twice)} = \binom{5}{2} \times 0.4^3 \times 0.6^2 \approx 0.35$$

which also answers the question I posed on page 29. Your chances are again a bit more complicated to compute. As both your guess and the actual outcome have binomial distributions, we need to use the law of total probability. If you guess 0, which happens with probability $0.6^5 \approx 0.08$, you are correct if the bus is never late, which also happens with probability 0.08. The first term in the law of total probability is $0.08 \times 0.08 = 0.08^2$. For the next term we need the binomial probability when $k = 1$, and this is $5 \times 0.4 \times 0.6^4 \approx 0.26$. This is the probability that you guess 1, and in that case, you are correct if the actual number is also 1. Thus, square this probability and add to the previous:

$0.08^2 + 0.26^2$. Continue in this way, compute and square each of the six binomial probabilities for k from 0 to 5, and then add them all. The result is the probability that you guess correctly and you may verify that it is about 0.25. Albert has been avenged.

Let us toss some more coins. If you toss four coins, the "typical" outcome is to get two heads and two tails. The probability of this outcome can be computed with the binomial distribution as

$$\text{P(two heads)} = \binom{4}{2} \times (1/2)^4 = 3/8$$

Likewise, the typical outcome when you toss six coins is three heads, when you toss eight coins, four heads, and so on. In general, how likely is it that you get the typical outcome, that is, equally many heads and tails when you toss an even number of coins? Suppose, thus, that you toss $2 \times n$ coins and ask for the probability to get n heads, in particular when n gets large. By the binomial distribution

$$\text{P}(n \text{ heads}) = \binom{2 \times n}{n} \times (1/2)^{2 \times n}$$

and it is not so easy to see where this heads for large n. There is, however, a nice approximation formula for factorials, called *Stirling's formula*, that comes in handy. This formula is quite technical, and I do not want to go into that kind of detail, so if you are interested, look up the formula on your own. It is pretty neat. Anyway, it turns out the the approximate probability to get equally many heads and tails is

$$\text{P(equally many heads and tails in } 2 \times n \text{ tosses)} \approx 1/\sqrt{n \times \pi}$$

You may wonder what on earth the number π is doing in there. Isn't that the ratio of the circumference and the diameter of a circle, the famous 3.14? Yes indeed, but as anybody who has studied mathematics knows, the number π tends to pop up in the most unexpected situations.[6] Better get used to it; you will see it again. Just for fun, let us try the formula for four coins, that is,

[6]Other seemingly non sequitur appearances of π are that if an integer is chosen at random, the probability that it is *square-free* (cannot be divided by any square such as 4, 9, ...) is $6/\pi^2$, and that if two integers are chosen at random, the probability that they are *relatively prime* (have no common divisors) is also $6/\pi^2$.

$n = 2$. The approximation gives

$$\text{P(two heads and two tails)} \approx 1/\sqrt{2 \times \pi} \approx 0.40$$

and the exact answer we got above was $3/8$, in decimal notation 0.375. Not bad. Of course, this is about as small an n as we can have and the approximation formula only gets better when n is large, which is also when the formula is useful. Note that the probability $1/\sqrt{n \times \pi}$ goes to 0 as n increases. Thus, the typical outcome is actually very unlikely. It is only typical in the sense of an average; if the $2 \times n$ coins are tossed over and over, the average number of heads will be near n, but it will not happen often that we get *exactly* n heads. More about this issue later in the book.

Let me finish with a sports example. Is it easier for the underdog to win the Super Bowl or the World Series? The difference is that the Super Bowl is a single game, but the World Series is played in best of seven games. Which benefits the weaker team? Let us ignore all practical complications such as home field advantage and all kinds of unpredictable events and simply assume that each game is won by the underdog with probability p, independently of other games. In order to win the World Series, four games must be won and if this goal is achieved in less than seven games, the remaining games are not played. This gives the following four different ways to win: win four straight games; win three of the first four games and the fifth; win three of the first five games and the sixth; and win three of the first six games and the seventh. In each of these cases, the last game must be won and this has probability p. Moreover, three games must be won among the first three, four, five, or six games, and in these cases, the number of games won has a binomial distribution with n equal to 3, 4, 5, and 6, respectively. The probability that the underdog wins can now by computed as

$$\text{P(underdog wins)} = \sum_{n=3}^{6} \binom{n}{3} \times p^3 \times (1-p)^{n-3} \times p$$

where we can try different values of p. In Table 1.2, the Super Bowl and World Series are compared for different winning probabilities for the underdog. Note that it is always easier for the underdog to win the Super Bowl than the World Series; the better team always benefits from playing more games. For example, a team that has a one in five shot to win a single game has only a 3.3% chance to win the World Series. We would expect more upsets in the

Table 1.2 Probability that the underdog wins the Super Bowl and the World Series

P(win single game)	50%	40%	30%	20%	10%
P(win Super Bowl)	50%	40%	30%	20%	10%
P(win World Series)	50%	29%	13%	3.3%	0.3%

Super Bowl than in the World Series, but I leave it to you to do the empirical investigation of historical championship data.

FINAL WORD

We are done with the introductory chapter. You are now armed with knowledge about probabilities and how to compute and interpret them. It's time to get to work.

2

Surprising Probabilities: When Intuition Struggles

BOYS, GIRLS, ACES, AND COLORED CARDS

Probability is notorious for problems that are easy to state but whose solutions can lead to confusion, heated disputes, and hurling of insults. Even though I have never personally witnessed a discussion of a probability problem turn into a fist fight, I would not consider it to be impossible. Most often the confusion stems from problems that are not well formulated and several different interpretations are possible. The high emotional level may have to do with the fact that probability problems often relate to everyday phenomena. It is in this way different from many branches of mathematics because you can often understand the problem and suggest solutions without having any formal training in probability. Let us start with a standard problem of this kind.

Your Aunt Jane calls again, this time to tell you that her new neighbors have two children and that at least one is a boy. What is the probability that the other is also a boy?

You discuss this problem with Albert and Betsy down at the pub. Your opinion is that one child is a boy and the gender of the other child is independent of this fact; thus, the probability is simply $1/2$. Albert, on the other hand, wants to be more methodical. He lists the sample space BB, BG, GB, and GG, and notes that the outcome GG is impossible. As we know that at least one child is a boy, we have either BG, GB, or BB, and as one out of the three has another boy, Albert suggests the probability $1/3$. Betsy, without motivation, suggests

1 (which causes Albert again to sigh into his pint), and Bob the bartender says 0 (he knows that Albert is a bad tipper anyway). Daisy the drunk suggests 0.73, which she thinks is a funny number. Who is right?

Actually it could be anybody. It depends on what assumptions we make and how Aunt Jane got her information. Let us start with Albert's solution. First of all, he assumes that the four outcomes are equally likely, which is reasonable. He claims that the information we have is "at least one boy" and thus arrives at the probability $1/3$. But how did Aunt Jane find out? Perhaps she asked the neighbors if they have any sons and got the answer "Yes." Then Albert's solution is correct. The information given allows us to rule out the outcome GG, and we are left with three equally likely possibilities.

Suppose that Aunt Jane later sees the mother taking a walk with a boy. Does this change anything? At first glance it does not seem to. After all, she already knew that there was a boy in the family and her observation just seems to confirm this. But observing a *particular* boy actually gives more information than being told "at least one boy," and we need the sample space to take this into account (this is similar to the situation we had on page 7 where we chose a girl from a three-children family). The sample space must identify which boy that is being seen, and with an asterisk denoting the boy walking with the mother, we get the equally likely outcomes

$$B^*B,\ BB^*,\ B^*G,\ GB^*$$

As two out of four outcomes have two boys, we conclude that the probability is $1/2$. Whenever the information is of the kind that a specific boy is identified, the probability is $1/2$. This is, for example, the case if Aunt Jane sees him, hears him answer the phone, or finds his lost report card in the street. The reason is that the outcome BB must then be split into two, taking into account that the identified boy can be either of the two. This is equivalent to saying that the gender of the other boy is independent of the one observed, so in this case your solution is correct. I would say that this is the most logical interpretation of Aunt Jane's statement if we are not given any specifics.

So what about Betsy and her probability 1? She could argue against our assumption that the four outcomes listed above are equally likely and for example suggest that the mother would never take a walk with her son and have to chase him through the mud puddles if she instead had a choice to take a leisurely stroll with her well-behaved daughter. This would give probability 0 to both B^*G and BG^* and probability $1/2$ to each of B^*B and BB^*. Obviously, if the observed is a boy under Betsy's assumption, the other child must be a boy as well.

Bartender Bob instead claims that a mother of two boys would never take one for a walk and leave the other at home with free hands to attack the walls with crayons, supervised only by the TV-watching father. No, according to Bartender Bob, if a boy is observed walking with the mother, you can be certain that the other child is a girl. Formally, Bartender Bob has assigned probability 0 to both B^*B and BB^* and probability $1/2$ to each of B^*G and GB^*.

In general, any four numbers between 0 and 1 that add up to one can be used as probabilities of the four outcomes and we can then come up with any number between 0 and 1 for the probability that the other child is a boy, including Daisy's 0.73. You can buy a fresh round of pints to celebrate that you are all correct (but not for Daisy, she's had enough). You may have your own opinion about what probability assignment is the most realistic, but aside from this, the example illustrates that in order to compute probabilities, you need to clearly specify under what assumptions you are working. In this case, depending on the assumptions, the probability could actually be anything from 0 to 1 and that is, well, kind of vague.

Here is a related problem. You deal a poker hand to Betsy and ask her, "Do you have any aces?" to which she answers, "Yes." What is the probability that she has more than one ace? You then ask, "Do you have the ace of spades?" and she again answers, "Yes." Now what is the probability that she has more than one ace?

Just like in the case of the children, there is a difference between the two pieces of information. And just like in the case of the children, there is more information in the case where something specific, in this case the ace of spades, has been identified. In the first case, we are asking for the probability of more than one ace given *at least one* ace; in the second, given one *particular* ace. To simplify things, consider a reduced deck, consisting of the four cards $\spadesuit$A, $\heartsuit$A, $\clubsuit$8, and $\diamondsuit$5, and suppose that you deal Betsy two cards. She thus has a reduced poker hand from a reduced deck where the six possible hands are

$$(\spadesuit A\ \heartsuit A),\ (\spadesuit A\ \clubsuit 8),\ (\spadesuit A\ \diamondsuit 5),\ (\heartsuit A\ \clubsuit 8),\ (\heartsuit A\ \diamondsuit 5),\ (\clubsuit 8\ \diamondsuit 5)$$

If she answers yes to the question "Do you have any aces?" then this excludes only one hand, the one with $\clubsuit$8 and $\diamondsuit$5. Of the remaining five, one has two aces and the conditional probability of two aces is $1/5$. If she answers yes to the question "Do you have the ace of spades?" then three hands are excluded, and the conditional probability of two aces is $1/3$.

I hope that you are convinced that the two conditional probabilities will be different also in the original problem with the full deck. It is just a bit more complicated to compute the probabilities. This problem is also described in *Lady Luck* where Warren Weaver attributes the problem and its solution to Martin Gardner, America's uncrowned King of Recreational Mathematics, who for 25 years, starting in 1956, ran the column "Mathematical Magic" in *Scientific American*, devoted to popularizing mathematics to a wide audience. Oddly enough, Mr. Weaver expresses some doubt that there really is a difference between "at least one ace" and "the ace of spades." In Mr. Weaver's own words:

> But is this reasoning of Mr. Gardner's really sound? Are the five hands (in the first case) and the three hands (in the second case) *equally probable*?

The answer is yes, and I think the simplified version with four cards explains this perfectly well.[1] The reason that Weaver was hesitant to accept Gardner's solution seems to be that he saw an analogy with another problem, also in *Lady Luck*, which we look at next.

In a box there are three cards. One is red on both sides, one is black on both sides, and one is red on one side and black on the other. Now draw one card at random and show one of its sides. Suppose that it is red. What is the probability that the other side is also red?

This should be easy. You draw at random so that the three cards are equally likely. If it has a red side, it is either the red/red card or the red/black card. The probability that it is the red/red card is therefore $1/2$. Plain and simple. But wait a minute. The same argument applies if the side that is shown is instead black. The probability that it is the black/black card is then also $1/2$. So regardless of what color we observe, the probability is $1/2$ that the other side has the same color. But this means that we get a "same-colored" card only half of the time even though there are two such cards out of three and we choose at random. Something is wrong here but what?

The problem lies in confusing unconditional and conditional probabilities. Initially, the red/red and the red/black cards are equally as likely to be picked;

[1]Gardner's formulation is in terms of bridge rather than poker. I do not know how to play bridge. The closest I have ever come was playing a simple type of whist with patients in a psychiatric clinic where I worked in the summers during my college years. I got discouraged when they told me that I played like an idiot.

that is, the unconditional probabilities for the two cards are the same. However, given that a red side is shown, the conditional probabilities of red/red and red/black are not equal anymore. Why not? This situation is very similar to the problem of the children above where we split up pairs to take into account which child was observed. Here, we must take into account which side of the card is shown. As each card has two sides, we get the following sample space with six equally likely outcomes where the asterisk indicates the side that is shown:

$$R^*R,\ RR^*,\ R^*B,\ RB^*,\ B^*B,\ BB^*$$

We can now easily compute any probability we want. In particular, it is easy to see the conditional probability of red/red given that the shown side is red is $2/3$ because three outcomes contain R^* and two of these are combined with an R. In contrast, the probability that it is the red/red card given that *at least* one side is red is $1/2$ because this condition also includes the outcome RB^*, but this is not the correct solution given the way I described the problem.

Warren Weaver saw an analogy between the poker problem and the colored cards problem. In the latter, at first the red/red and the red/black cards are equally likely, but after eliminating the black/black card, they are not equally likely anymore. This insight led him to question whether the elimination of some of the poker hands leaves the remaining hands equally likely. But as we have just seen, a complete description of the colored cards problem requires not three but six equally likely outcomes, and if the observed side is red, we eliminate three outcomes out of six, not one out of three. As a matter of fact, the reduced poker problem with a deck of four cards also has six equally likely outcomes so there is indeed a close analogy between the problems but not in the way that Mr. Weaver intended.

It is very easy to check these probabilities empirically (preferably on a computer). In the poker problem, use the four cards mentioned above. Then repeatedly shuffle and deal two cards. Disregard the times you get no aces, and count the number of times you get at least one ace. Call this number N. Simultaneously, also count the number of times you get the ace of spades, call this number A, and the number of times you get two aces, call this number T. After a while, T/N should be approximately $1/5$ and T/A should be approximately $1/3$.

GOATS AND GLOATS

The "Monty Hall problem" is probably the most famous of all probability problems. It created a drama that involved a TV show, the world's smartest person, several rude mathematicians, one car, two goats, and countless lunch discussions all over the world. The TV show in question was the game show "Let's Make a Deal" with host Monty Hall, and the problem is as follows: You are given the choice of three doors. Behind one door is a car; behind the others are goats. You pick a door without opening it, and the host opens another door and reveals a goat. He then gives you the choice to either open your door and keep what is behind it, or switch to the remaining closed door and take what is there. Is it to your advantage to switch?

At first, it may seem that it does not matter because there are two doors, one of which hides the car. But because Monty knows where the car is, he will *always show you a goat*, which means that by switching, you win the car whenever your initial choice was a door that hid a goat, and the probability of this is $2/3$. Thus, it is to your advantage to switch. As this problem has been discussed, debated, and explained *ad nauseam*, I will leave it at this. There are plenty of books and websites that offer various explanations and extensions, and several websites where the game can be played. Google away. I will instead attempt to explain where the controversy might stem from but, first, a brief history.

The problem gained fame after it appeared in the column "Ask Marilyn" in *Parade Magazine* in 1991. The column is written by Marilyn vos Savant who in the late 1980s was listed in the *Guinness Book of World Records* as having the highest IQ (228) ever recorded.[2] After Marilyn had explained why it is to your advantage to switch, a slew of mathematicians attacked her, claiming that she was wrong. I suppose they felt a great deal of schadenfreude when they thought that this woman, more intelligent and certainly much more famous than themselves, had erred on such a simple problem. Some referred to blind trust in authority:

> I am in shock that after being corrected by at least three mathematicians, you still do not see your mistake.

[2]The category of highest IQ has since been discontinued, and Marilyn has been retired to the Guinness Hall of Fame. Test score ceilings were lowered so that IQs as high as Marilyn's could no longer be recorded. It is questionable how meaningful such high scores are, but it must be nice to be famous for being the world's most intelligent person. Just think of all the opportunities it gives you like...having your own column in *Parade Magazine*.

> How many irate mathematicians are needed to get you to change your mind?
>
> If all those Ph.D.s were wrong, the country would be in very serious trouble.

Some were quite rude:

> You blew it, and you blew it big!...You seem to have difficulty grasping the basic principle at work here...There is enough mathematical illiteracy in this country, and we don't need the world's highest IQ propagating more. Shame!
>
> May I suggest that you obtain and refer to a standard textbook on probability.
>
> I am sure you will receive many letters from high school and college students. Perhaps you should keep a few addresses for help with future columns.

And there were those who tried a more, ehm, understanding approach:

> Maybe women look at math problems differently than men.

And the story ends with Marilyn being correct and all the irate mathematicians proven wrong and the gloaters receiving the gloat.[3] One example: British statistician Brian Everitt wrote a nice book called *Chance Rules: An Informal Guide to Probability, Risk, and Statistics* where he describes the Monty Hall problem and comments that one of the named mathematicians is still "wiping egg off his face." Come on now, let's not kick him when he's down. One may object to the harsh tone in some of the letters, but I will say a few words in defense of these mathematicians. I may be overly benevolent, but they could all have instinctively been thinking about the conditional probability given that a door *randomly chosen* by Monty reveals a goat. This is different. Let us look at it and assume throughout that you have adopted the strategy always to switch doors.

If both you and Monty choose doors at random, there are three equally likely outcomes (your choice first): car/goat, goat/car, and goat/goat. Why equally likely? Well, you have probability $1/3$ to choose the car and then he

[3]The dissenting mathematicians were in good company though; Paul Erdős (1913–1996), one of the greatest mathematicians of the twentieth century, allegedly also had difficulty in accepting that one should switch. In the Erdős biography, *The Man Who Loved Only Numbers*, Paul Hoffman tells the story about Erdős and Monty Hall and how the great mathematician was convinced that one should switch doors only after one of his friends showed him a computer simulation of the problem.

always gets a goat. You have probability $2/3$ to choose a goat, and he has then (conditional) probability $1/2$ to choose the car and $1/2$ to choose a goat, which gives probability $1/3$ to both goat/car and goat/goat. If his random choice reveals a goat, the outcome goat/car is ruled out and the equally likely outcomes car/goat and goat/goat remain. In one of them you win, and in the other you lose, so your probability to win is $1/2$. Notice that it is here possible for Monty to reveal the car; he just didn't do it in this case. You can think of a situation where the game is played like this repeatedly and the cases when Monty reveals the car are disregarded. This happens about one third of the time, and in the remaining cases, you win half of the time. In the game show context, the natural interpretation is of course that Monty, who knows where the car is, always shows you a goat. In that case, the outcome goat/car has probability 0 and your winning outcome goat/goat has probability $2/3$.

One psychological aspect of the Monty Hall problem is that it would feel pretty bad to switch doors and find a goat. It would feel like you had the car but traded it for a goat. But if you stick to your initial choice and find a goat, at least you stood firm, you were not indecisive and wishy-washy and there is a certain appeal to that. Perhaps we should try to quantify this "frustration factor" and weigh it into the solution of the problem.

This type of probability problem did not originate with Marilyn and Monty. An older problem, which is in essence the same, is the "three prisoner's paradox," described among others by Martin Gardner in the 1950s. Here is my own suggested biblical version: King Nebuchadnezzar decides to randomly spare either of Shadrach, Meshach, or Abednego from being thrown into the blazing furnace. Abednego, slightly more apprehensive of the flames than his friends, asks if he is the one spared. Nebuchadnezzar does not want to answer this, but he does tell Abednego that Shadrach will not be spared. The answer relieves Abednego somewhat because he figures that his chance to be spared has now increased from $1/3$ to $1/2$. I leave it to you to investigate this, find the similarity with the Monty Hall problem, and take away from Abednego the little hope he had.

HAPPY BIRTHDAY

In many problems in probability, it is difficult to have an intuitive grasp of what the correct answer should be, even if the problem is simple to state and refers to something familiar. Probability problems are in this way different from other types of everyday math problems. For example, as we are constantly involved in various scheduled activities (classes, work hours, travel, etc.), we are pretty good at getting ballpark figures regarding time. Even if we have never driven between New York and Chicago, it wouldn't be too hard to approximate how long it would take. And certainly, if somebody suggested that it would take about two weeks, we would find this ridiculous. Likewise, we are fairly good with physical measures of length, area, and volume. You don't have enormous problems estimating the tip on a restaurant bill. And you might be reasonably certain that if you want to move the furniture out of your house in an afternoon, you would have to ask five or six friends to help you. But how about this: How many people do you need to ask to be at least 50% certain that at least two of them have the same birthday?

You may know the solution because this is a fairly well-known problem. But even if you do, can you honestly say that you have a good feeling for it? We certainly do not have a lot of empirical experience; rarely do we examine groups of people for coinciding birthdays. Let us look at the solution and make a few simplifying assumptions. First, we exclude February 29th to assume that a year has 365 days, and second, we assume that all days are equally likely birthdays for a randomly chosen person. We can now employ Trick Number One and compute the probability that everybody has a *different* birthday, and then subtract this from one. Let us start with two people. The first can have any birthday; the second must avoid this day, which has probability $364/365$. The probability that two people share a birthday is thus $1 - 364/365 \approx 0.003$.

Add another person. Her/his birthday must then avoid both previously taken birthdays, which has probability $363/365$. The probability that three people have different birthdays is, therefore, $364/365 \times 363/365$ (formally we are using the multiplication rule for conditional probabilities from page 20 here), and the probability that there is some common birthday in a group of three is

$$\text{P(some common birthday)} = 1 - \frac{364}{365} \times \frac{363}{365} \approx 0.01$$

and by proceeding in this way, we realize that the probability that there is some common birthday in a group of n people is

$$\text{P(some common birthday)} = 1 - \frac{364}{365} \times \frac{363}{365} \times \cdots \times \frac{(366 - n)}{365}$$

This number increases surprisingly quickly as n increases. For n equal to 4, 5, 6, and 7, we get 0.02, 0.03, 0.04, and 0.06, respectively. The probability exceeds 0.1 already at ten people; for $n = 22$, it is 0.48, and for $n = 23$, the probability of some common birthday is 0.51. Thus, only 23 people are needed to be at least 50% certain that there is some common birthday! Quite astonishing, don't you think? To verify this result empirically I looked up the NFL rosters for 2006 and checked the 23 first players on each team (I've been working hard for you dear reader). Out of the 32 teams, I found 17 that had at least two players that shared birthdays. Of course we don't expect to be that close every time, but I'm glad that already my first attempt gave such nice agreement between theory and observations. Some probabilities of a shared birthday in groups of various sizes are given in Table 2.1.

Table 2.1 Probabilities (as percentages) of a shared birthday

Number of people	5	10	22	23	30	50
P(shared birthday)	2.7%	11.7%	47.6%	50.7%	70.6%	97.0%

It is interesting to note that 23 is far from 183, which is about half of 365. I would think that many people's first guess for the number of people needed for a 50% chance of a common birthday would be around 183. I have sometimes in the classroom asked for the probability that there is at least one common birthday in a group of 100 people. I draw a line, mark 0 and 1, start sliding a pointer from 0 toward 1 and ask the students to raise their hands when they think I hit the right probability. Indeed, most hands go up somewhere between $1/4$ and $1/3$ (both in Sweden and in the United States in case anybody thought there would be cultural differences). The correct probability is 0.9999997.

Suppose now that I sit down with a bunch of school catalogs from my elementary school years in Sweden. There are ten classes with 20–25 kids in each class. I check how many classes that have somebody who shares my birthday. According to the result above, I would expect to find this in about five classes, give or take a few. Too my disappointment, I find none. Very bad luck?

No, this result was actually to be expected. It is different to ask for the probability that somebody shares somebody else's birthday and that somebody shares a *particular* birthday. In the computation in the first case, each new person must avoid all previously taken birthdays as we saw above. In the second case, it is enough to avoid the particular birthday, and each time the probability of this is $364/365$. In a group of 23 people (excluding me), the probability that somebody shares my birthday can then be computed with Trick Number One as

$$\text{P(somebody shares my birthday)} = 1 - \text{P(nobody shares my birthday)}$$
$$= 1 - (364/365)^{23} \approx 0.06$$

which is indeed different from 0.51. How many people are then needed for the probability that somebody shares my birthday to exceed 0.5? Replacing 23 in the exponent with 252 gives a probability just below 0.5, and with 253, the probability is just above 0.5. Thus, me excluded, there needs to be 253 people, substantially more than 23.

One can note that the numbers 23 and 253 are both far from 183, but it turns out that there is an interesting connection between them. Some understanding of the problem can be gained by thinking in terms of the number of *possibilities* to match birthdays. Let me explain. If we first consider my birthday, we know that if it is compared with 253 other people's birthdays, chances are 50–50 that there is a match. Thus, we have formed 253 *pairs* of birthdays and looked for matches within these pairs. If we instead want to find any match, not necessarily with me, this means that we compare everybody with everybody else and form all possible pairs. In a group of 23 people, how many pairs are there? Remember your combinatorics from the previous chapter to realize that the number of pairs is $\binom{23}{2}$, which equals $23 \times 22/2$, which equals...drum roll here...253! Thus, with 23 people, we have 253 chances to find a pair that matches, and the probability of this is about 0.5.

Although the argument above is illuminating and useful, it involves a little bit of cheating. Recall how we computed the probability of somebody sharing my birthday as $1 - (364/365)^{253}$. The repeated multiplication of $364/365$ by itself is done because birthdays of different people are independent. Each time a new pair is formed, this is done by combining me with somebody new and such pairs are independent in the sense that if I get a match with one person, this does not tell anything about whether I will get a match with the next (unless they are chosen in some particularly odd way, for example, at a convention for people born on February 29th). However, when we form the 253 pairs from the group of 23, these pairs are not independent. For example, I have the same birthday as Freddie Mercury, and if I find out that he shares birthday with Jesse James (the outlaw), then I know that Jesse and I share birthdays, too. But many of the pairs are independent of many other pairs. I and Freddie share birthdays, and if I am told that Madonna and Charles Bukowski share birthdays (they do), then this does not say anything about whether Madonna and I share birthdays (we don't). There are enough such independent pairs that the argument is valid as a good approximation (and as we saw, it does give the correct relation between 23 and 253). Thus, in a group of n people, the probability that at least two have the same birthday is approximately

$$\text{P(some shared birthday)} \approx 1 - \left(\frac{364}{365}\right)^{\binom{n}{2}}$$

which gives you a quicker way to compute the probability.

We assumed that all birthdays are equally likely, but in real life, they are not. However, our assumption of equally likely birthdays makes it *more* difficult to share birthdays, so in reality, the already surprisingly low number 23 is even lower. I will not give the complete mathematical proof of this but give two different arguments, one intuitive and one mathematical. Hopefully at least one of these will convince you.

Intuitive argument: Consider some extreme deviation from equally likely birthdays. Suppose, for example, that everybody was born on January 1. Then how many people do you need to be at least 50% certain that there is a shared birthday? Obviously, two is enough. A little less extreme: Suppose everybody was born in January. Then there are 31 days to choose from rather than 365, and the probability that there is some shared birthday among n people is

$$\text{P(some shared birthday)} = 1 - \frac{30}{31} \times \frac{29}{31} \times \cdots \times \frac{(32-n)}{31}$$

and it is readily checked that this exceeds 0.5 already at $n = 7$. Thus, only seven people are needed. With 23 people, you can be almost certain that there is a shared birthday. Now, these are just some examples of skewed birthday distributions, but whichever you come up with, the number of people needed for a 50% chance of a shared birthday will never be more than 23.

Mathematical argument: To simplify things considerably, divide the year into "winter half" and "summer half" and consider "birth-halfyears" instead of birthdays. Suppose that you have two people. If they are equally likely to be born either half, the probability that they are born the same half is simply $1/2$ (first person can be born any half, and second person must be born the same half as the first). Now suppose that each person is born in the winter half with probability p and in the summer half with probability $1 - p$, independently of each other. The probability that both people are born in the winter half is then $p \times p$, and the probability that they are both born in the summer half is $(1-p) \times (1-p)$. Thus, the probability that they are born in the same half is $p^2+(1-p)^2$, and it is easily checked that this is smallest when $p = 1/2$ (either by trying some values of p and be convinced, by plotting the probability as a function of p, or by using techniques from calculus to find the minimum of a function). Thus, the probability that there is a shared "birth-halfyear" for two people is *at least* $1/2$. In the same way, the probability that there is a shared birthday among 23 people is at least $1/2$, whatever the true distribution of birthdays may be.

The birthday problem can be extended in many ways. For example, how many people are needed to be at least 50% certain that at least *three* people share birthdays? About 82. For another example, how many people are needed to be at least 50% certain that at least two have birthdays not more than one day apart? Only 14, but now I think that nothing will surprise you anymore, so we leave the birthday problem and turn to something different. We will return to the birthday problem in the gambling chapter where I will describe how you can use it to make money. Yes, really.

TYPICAL ATYPICALITIES

Situations where randomness is present are often summarized by use of the terms "average" or "typical." In 1940, The Leathers family of Clarendon, Texas was chosen as the "typical American family" for New York's World Fair. The "average American" carries more than $8,000 in credit card debt. Unless you know exactly what they mean, such statements can be very misleading. We will return in more detail to averages in Chapter 8 but will briefly touch on it here and illustrate how what might be considered typical may actually be unlikely.

Consider a randomly chosen family with four children. What is the most likely number of daughters? Of sons? What is the most likely gender split: all children of the same gender (0–4), one child of one gender and three of the other (1–3), or two children of each gender (2–2)?

It seems that the answers should be 2, 2, and 2–2, respectively. Let us check. The number of daughters has a binomial distribution with $n = 4$ and $p = 1/2$. The probabilities of 0, 1, 2, 3, and 4 daughters can be computed with the formula on page 39 and are easily found to be 6.25%, 25%, 47.5%, 25%, and 6.25%, respectively. Indeed, the most likely number of daughters is

2. In the same way, the most likely number of sons is also 2. And this answers the last question automatically; the most likely split is 2–2. Correct?

Of course not, or it wouldn't be in this chapter. The most likely split is actually 1–3, which has probability 50%. This result is not so mysterious because it includes one son and three daughters, which has probability 25%, and one daughter and three sons, which has probability 25% also. Add them to get 50%. It is a little bit odd though if we express it like this: The typical four-children family has two girls, two boys, and a gender split of 1–3! The oddity comes from the fact that the "typical family" is of course not a very clearly defined concept.

A similar situation arises in card games where it might be more of a surprise. In bridge, all 52 cards are dealt so that the four players get 13 cards each. What is the most likely distribution of the numbers of cards in the four suits? As you have 13 cards and 4 equally likely suits, on average you should get 3.25 cards in each suit. Rounding to the nearest integer combination, you get 4–3–3–3. So, in some sense, this should be the "typical" distribution. But is it the most likely? No, by computing the probabilities of this and other possible combinations, it turns out that the distribution 4–4–3–2 is the most likely. It has probability 22%, compared with only 11% for 4–3–3–3. As a matter of fact, several other distributions are more likely than 4–3–3–3; see Table 2.2.

Table 2.2 The six most common suit distributions in bridge

Suit distribution	Probability
4–4–3–2	21.6%
5–3–3–2	15.5%
5–4–3–1	12.9%
5–4–2–2	10.6%
4–3–3–3	10.5%
6–3–2–2	5.6%

This result seems counterintuitive, but in analogy with the gender distributions above, remember that we do not distinguish between, for example, "5 hearts, 3 clubs, 3 diamonds, and 2 spades" and "5 diamonds, 3 hearts, 3 clubs, and 2 spades." However, I would think that such a distinction is of less interest in bridge than the numbers alone (but remember footnote 1 on page 48).

You might be curious about how to compute the probability of a particular suit distribution. It all boils down to combinatorics, and let us illustrate this for the case 4–3–3–3. As a bridge hand consists of 13 cards chosen randomly from a total of 52, there are $\binom{52}{13}$ different bridge hands. Next, we must find the number of such hands that give the suit distribution 4–3–3–3, and there are plenty of them. Let us start with those that have 4 hearts and three in each of the other suits. The 4 hearts can be chosen from the 13 hearts in $\binom{13}{4}$ ways. For each such choice, the three clubs can then be chosen in $\binom{13}{3}$ ways. For each such combination of four hearts and three clubs, the three diamonds can be chosen in $\binom{13}{3}$ ways, and finally, the spades can also be chosen in $\binom{13}{3}$ ways. The multiplication principle then tells you to multiply these four numbers to get the total number of hands that give the distribution 4–3–3–3 with four hearts. Equally as many hands have the distribution 4–3–3–3 with four clubs, four diamonds, or four spades, and we finally get

$$P(\text{suit distribution } 4\text{–}3\text{–}3\text{–}3) = \frac{4 \times \binom{13}{4} \times \binom{13}{3} \times \binom{13}{3} \times \binom{13}{3}}{\binom{52}{13}} \approx 0.11$$

As an exercise, you may convince yourself that the probability of the most likely distribution 4–4–3–2 is

$$P(4\text{–}4\text{–}3\text{–}2) = \frac{\binom{4}{2} \times 2 \times \binom{13}{4} \times \binom{13}{4} \times \binom{13}{3} \times \binom{13}{2}}{\binom{52}{13}} \approx 0.22$$

The *least* likely suit distributions are 13–0–0–0, which can occur in four ways, followed by 12–1–0–0, which can occur in $13 \times 13 \times 4 \times 3 = 2{,}028$ ways (explain this!).

Let us now consider the game of darts. In case you didn't know it, a dart board is a circular board divided into 20 sectors (pie slices) numbered 1–20 (not in order). In the center of the board is the "bulls eye," and the board has two concentric rings, each about 0.3 inches wide. The inner of the two is the "treble ring," where the score triples, and the outer is the "double ring," where the score doubles. That's about all we need to know.

Now suppose that you repeatedly throw darts at random on such a board.[4] Your darts will be uniformly spread over the board; you are equally likely to

[4]Needless to say, dart players do not throw at random at all, a fact that has even been used in court. In 1908, Jim Garside, an English inn keeper, was charged with allowing customers to bet on a game of chance: darts. Garside called in local champion William "Bigfoot" Anakin

hit 6 and 11, equally likely to hit double 20 and double 3, and so on. Clearly, the average position will be the bull's eye but in no other way is this "typical."

There is another interesting thing about the dart board. As each position can be described by the angle to the horizontal axis and the distance to the origin (the very center of the bull's eye), we could let a computer simulate a randomly thrown dart by first choosing an angle at random between 0° and 360°, and then choosing a radius at random between 0 and 9 inches (which is roughly the radius of a dart board). However, if we do this, it turns out that the darts tend to be more concentrated toward the center of the board. Why?

The answer is that when a dart is thrown at random, larger distances to the bull's eye are more likely to occur than shorter. For example, it is easier to hit the double ring than the treble ring because, even though they are equally wide, the double ring has a larger area being farther from the center. There is also a ring around the bull's eye, also of the same width, and this is even more difficult to hit. Therefore, choosing a distance at random in the sense that all distances are equally likely, will *not* describe a randomly thrown dart. It can be shown that the average distance between randomly thrown darts and the center of the board is not equal to half the radius of the board, but rather $2/3$ of that radius. The mathematics needed to compute this average distance is more than I want to get into in this book, but it is easy to verify experimentally (with computer aid, preferably).

A similar oddity arises if you want to choose a point at random on the surface of the earth by choosing a latitude at random between 90°N and 90°S and a longitude at random between 180°W and 180°E. If you do this repeatedly, you will notice that your points tend to have a pretty high density near the poles and be much sparser near the equator. The explanation here is that whereas longitudes are of equal length, latitudes are not. Indeed, the two extremes 90°N and 90°S correspond to only one point each, the two poles, but the 0° latitude is the entire equator. If you choose your vacation destination with this method, you are far more likely to end up in Spitsbergen than at the Raffles Hotel in Singapore.

as a witness. Bigfoot showed that whatever number the court mentioned, he could hit on the board. Garside was acquitted as the court found darts to be a game not of chance but of skill.

STRATEGIES, SHOPPING, AND SPAGHETTI WESTERNS

Many decisions in life are made in the presence of randomness. And for many such decisions, trying to use whatever information you have at hand to make the best decision possible is important. Many probability problems involve how to choose the best strategy, both recreational problems and problems from real life. In many cases, real-life problems of this nature can be "purified" just like many probability problems can be described as problems about tossing coins or rolling dice. As this book is intended mostly for your entertainment, I will focus on the more playful side of strategies and decision making, but bear in mind that even serious matters such as stock markets, war, and marriage are often described in terms of puzzles and games.

The online store Amazon.com has a feature called the "Gold box." When you enter, you are presented ten special offers of various things that can be anything from books and DVDs to kitchenware and the "Panasonic ER411NC nose and ear hair groomer" (battery not included). The items are presented one at a time, and each time you must decide whether to buy it or skip it and proceed to the next. Once an item has been skipped, it is gone and cannot be retrieved. The rules used to be that once you decided to buy something, you would not get to see the rest of the offers, but now you can buy as many as you want; you just cannot backtrack and change your order. Suppose that you have decided to buy one of the items, no more and no fewer. What strategy should you employ to maximize the chance of "winning," that is, getting the best offer?

With "best offer" I mean that if you saw all ten offers, you could rank them from best to worst and take the best. The problem now is that you only see them one by one and have to decide on the spot whether to take or pass. What to do? Suppose that you use the impatient strategy of always taking the first offer. Assuming that the offers are presented in random order, you then have a 10% chance that the best offer is the first. If you instead use the hesitant strategy of looking at all offers, then you must take the last and again have a 10% chance. Try a random strategy, and choose a number between one and ten at random and take that offer. Still a 10% chance. Can you do better?

Certainly. Let the first five offers pass without taking any, and remember the best among them. Call it your target. Continue to view offers, and take the first that is better than the target. If this never happens, you are forced to take the last offer; the target was then the best offer and you missed it. With this strategy, you have a better than 25% chance to win. The reason is that you will always succeed if the second best offer is among the first five and the best offer is among the last five. The probability that the second best is among the first five is simply $5/10 = 1/2$ (it is equally likely to be in any of the ten positions). Given that the second best is among the first five, the conditional probability that the best is among the last five is $5/9$ (it is equally likely to be in any of the nine remaining positions). The probability that both events occur is, by the multiplication rule for conditional probabilities

$$\text{P(second best among first five and best among last five)} = 1/2 \times 5/9$$
$$= 5/18 \approx 0.28$$

which is certainly more than 0.25. Also notice that the true probability to win is even larger because what we described above is not the *only* way in which you can win. You also win, for example, if the best offer is number six and the second best is any of seven to ten. There are other cases as well. The best offer, of course, has to be among the last five, but as long as it is not preceded by another offer that beats the target, you win. You could find the exact probability to win by listing all possible orders of the offers, but this is tedious. You can do it on your own for a smaller number, for example, four offers. There are then $4! = 24$ ways to list the offers in order, and you can count in how many of these you win, using the "pass first half" strategy (for example, with the offers ranked from 1 to 4, 1 being the best, you win in the cases 4231, 3412, and a bunch of others). You should get the probability $10/24$.

Suppose now that instead of 10, there are 100 offers. You use the same strategy: Let the target be the best among the first 50 offers, and then choose the first offer that beats the target. One case in which you will definitely win is when the target is the second best and the best is among the last 50. The probability of this is $50/100 \times 50/99 = 25/99$, which is slightly above $1/4$ (which equals $25/100$). And again, there are other cases and your chance of picking the best offer is still at least 25%. You now realize that this result holds *regardless of how many offers there are*. If there are a million offers, find the target among the first half million and then take the first that beats the target. Probability of winning: at least 25%. Not bad and quite surprising but still not the best you can do.

It can be shown that an even better strategy is to choose your target among the first 37% of the offers and then choose the first that beats the target. Your chance of winning is then 37%. This strategy is actually optimal; no other strategy, no matter how complicated, gives you a higher chance of finding the best offer. The number 37% is an approximation that becomes more and more accurate as the number of offers grow. In the initial Gold box problem, 37% of 10 is 3.7, so you must round this to 4.

This problem is a classic in the probability literature and is often stated in terms of a princess trying to choose the best among a succession of suitors, or a prince trying to choose the bride with the largest dowry. I decided to lift this probability problem from the world of fairy tales into the crass reality of Internet shopping and to give an example of a situation that I have actually encountered. By the way, I passed on the nose and ear hair groomer.

Let us now escape the mundanity of everyday life and instead enter the Spaghetti Western world of Sergio Leone. In the final scene of the classic 1966 movie *The Good, the Bad, and the Ugly*, the three title characters, also known as "Blondie," "Angel Eyes," and "Tuco," stand in a cemetery, guns in hands, ready to shoot. Let us interfere slightly with the script and assume that Blondie always hits his target, Angel Eyes hits 90% of the time, and Tuco 50% of the time. Let us also suppose that they take turns in shooting, that whomever is shot at shoots next (unless he is hit), and that Tuco starts. What strategy maximizes his probability of survival?

First of all, it is a really bad idea to try to kill Angel Eyes because if he succeeds, Tuco is a dead man. Trying to kill Blondie might be a little better because if he succeeds, there is still a small chance that Tuco wins a shoot-out with Angel Eyes. Still, it is better if he misses since Blondie will then kill Angel Eyes who is a better shot than Tuco, and Tuco gets one last chance to kill Blondie. As it is better for Tuco to miss his first shot regardless of who he aims at, the conclusion is that he should miss on purpose! In this way, one of the two better shots will be eliminated by the other and Tuco only has to survive the final shoot-out. Who should he aim at?

If he aims at Blondie, the sequence of events is decided: Blondie kills Angel Eyes and Tuco gets one last chance to kill Blondie, which renders Tuco a survival probability of 50%. Even better is to aim at Angel Eyes and miss. If Angel Eyes misses Blondie in his next shot, which has probability 10%, the scenario from above is repeated and Tuco has probability 50% to survive in this case. If Angel Eyes kills Blondie, Tuco gets the final shoot-out with Angel Eyes so let us compute the probability that Tuco wins this if he gets to start.

To make things simpler, let us assume an infinite supply of bullets (hey, it's a Clint Eastwood movie we're talking about here). How can Tuco win? He can hit with the first shot. He can miss the first shot, Angel Eyes misses his first shot, and Tuco hits with his second. And so on. Just like Ann in the less fatal tennis example on page 29, Tuco can win in an infinite number of ways. And just like in the tennis example, we can solve this problem by obtaining an equation for the unknown probability that Tuco wins. There are two cases to consider:

I: Tuco hits with his first shot

II: Both Tuco and Angel Eyes miss their first shots

(for obvious reasons we do not have to consider the case when Tuco misses and Angel Eyes hits). The probability that Tuco wins is

$$P(\text{Tuco wins}) = P(\text{Tuco wins in I}) \times P(\text{I}) + P(\text{Tuco wins in II}) \times P(\text{II})$$

where $P(\text{Tuco wins in I}) = 1$, $P(\text{I}) = 0.5$, and $P(\text{II}) = 0.5 \times 0.1 = 0.05$. In case II, the shoot-out starts over and the two probabilities P(Tuco wins) and P(Tuco wins in II) are equal but unknown. Denoting this unknown number by p gives the equation

$$p = 0.5 + 0.05 \times p$$

which has the solution $p = 0.5/0.95 \approx 0.53$. Remember that this was in the case when Angel Eyes had succeeded to kill Blondie, which has probability 0.9. In the other case, with probability 0.1, Angel Eyes misses Blondie, which gives Tuco probability 0.5 of survival. All taken together, the law of total probability gives the final answer:

$$P(\text{Tuco survives}) = 0.9 \times \frac{0.5}{0.95} + 0.1 \times 0.5 \approx 0.52$$

so with his sneaky strategy to pretend to aim at Angel Eyes but miss on purpose, Tuco has a 52% survival chance. When Fredric Mosteller presented a similar problem in his 1965 book *Fifty Challenging Problems in Probability*, he expressed some worry over the possibly unethical dueling conduct to miss on purpose. In the case of Tuco, we can safely disregard any ethical considerations.

THE BRITISH SNOB AND I

Here is another problem in which we can use a method similar to the one we used to find Tuco's strategy above. Suppose that a certain type of cells are equally as likely to divide as they are to die. If you start a population from one such cell, what is the probability that the population eventually goes extinct?

Not so easy. The population can go extinct immediately, if the first cell fails to divide. This has probability $1/2$. If the first cell succeeds to divide, there are now two cells. The population can then go extinct in the next generation if both of these fail to divide, which has probability $1/2 \times 1/2 = 1/4$. It is also possible that both succeed to divide or that one fails and the other succeeds. As you can imagine, it quickly becomes complicated to compute probabilities. If there are some successful divisions in the beginning, we can end up with a lot of cells. And if we have a bunch of cells, we expect on average half to divide and half to die, which leaves on average the same number of cells in the next generation so the population tends to stay on average constant. This is only on average, though; there are of course plenty of random fluctuation in the actual population size. It is hard to imagine what will happen, but let us use the law of total probability in a clever way.

Let E be the event that the population eventually goes extinct, let S be the event that the first cell succeeds to divide, and let F be the event that it fails. The law of total probability then gives the following expression:

$$\text{P(E)} = \text{P(E given S)} \times \text{P(S)} + \text{P(E given F)} \times \text{P(F)}$$

where three probabilities are easy: $\text{P(S)} = \text{P(F)} = 1/2$ and $\text{P(E given F)} = 1$ because if the first cell fails to divide, the population goes extinct immediately. It remains to compute P(E given S), the probability of eventual extinction if the first cell succeeds to divide. But this means that we restart the population from two cells instead of one because each daughter cell starts a subpopulation (or lineage). For the entire population to go extinct, both subpopulations must go extinct. We have brought the problem back to where we started but with two initial cells instead of one. Let $p = \text{P(E)}$, the probability of eventual extinction. If we assume that the two subpopulations develop independently of each other, the probability that both go extinct is $p \times p = p^2$, and we thus have $\text{P(E)} = p$ and $\text{P(E given S)} = p^2$, which gives the equation

$$p = p^2 \times 1/2 + 1 \times 1/2$$

which is a so-called quadratic equation. You may know how to solve such equations in general, and if you do, you will get the solution $p = 1$. Extinction is inevitable! If you are not familiar with solutions to quadratic equations, you can insert $p = 1$ and see that it works (there are no other solutions in this case, although quadratic equations in general have two solutions). This is pretty unexpected. Even though the population stays constant on average, the random fluctuations will sooner or later cause it to go extinct.

The cell population is a simple case of something that probabilists call a *branching process*. The first people to study such processes were Frenchman I. J. Bienaymé and Englishmen Sir Francis Galton and Henry Watson. Sir Francis Galton who lived between 1822 and 1911 was an interesting figure. A man of independent means (in other words, he never worked for a living), he spent his life as a devoted amateur scientist. He studied meteorology, genetics, psychology, geography, tropical exploration, and statistics. His most famous contribution is probably the use of fingerprinting, a procedure certainly much more famous than the man himself. Galton is also far less famous than his cousin Charles Darwin. Sir Francis was obsessed with data collection and classification, especially of people. In his pockets he used to carry cards in which he poked holes to record the number of attractive and unattractive women he saw on the streets. His results led to a British "beauty map" that I have never seen referenced in any travel guide. He also investigated the "efficacy of prayer" and measured heights, weights, and several other traits of fathers and sons that led him to introduce the statistical concepts of *regression* and *correlation*, both of which I will explain later in the book. Among Galton's more dubious scientific contributions is the introduction of *eugenics*, the science of "bettering" the human race by selective breeding.[5] Although he would certainly have objected to certain later twentieth-century ideologies and their interpretation of eugenics, there might lie the reason that Galton has not become as acclaimed as his other scientific contributions seem to warrant.

Being the British snob that he was, Sir Francis worried about extinction of British nobility. And having the quantitative mind that he had, he posed this issue as a probability problem in a scientific journal. Galton's problem

[5]One supporter of eugenics was telephone inventor Alexander Graham Bell who studied people at the deaf colony on Martha's Vineyard. He concluded that deafness was hereditary and proposed a law against marriage between deaf people. He had nothing against deaf people though; both his mother and his wife were deaf.

was similar to the one about the cells above, one difference of course being that humans can have more than one offspring, which makes the problem more complicated. Galton did not solve the problem; this was done by his friend Henry W. Watson, reverend, mathematician, and mountaineer. Watson attacked the problem with mathematical methods that are still used today by probabilists who study branching processes. I am one of them. Most of my research in probability has been in this field, mainly applied to cell and molecular biology where I have investigated models for accumulation of mutations and for the progressive shortening of chromosomes that takes place as cells undergo successive divisions. The closest I have ever come to British nobility is a 2001 article about longevity among people in Massachusetts.

Let us take out the old deck of cards again. I will ask you to play the following game: I shuffle the deck and start dealing the cards face up, one by one. At any point you can say, "Stop! The next card is red" and if you are correct, you win. You must say it at some point, so if I have dealt 51 cards and you have not stopped me yet, you must guess that the last card is red. Other than that, you are free to use any strategy you wish. What is your best strategy, and what is then your probability to win?

If you guess that already the first card is red before I have started dealing, you have a 50–50 chance to win. Can you do better? If you let me deal the first card and this is black, you could stop me and have probability $26/51$ or about 51% that the next card is red. Then again, it could of course also happen that the first card is red, and you are then at a disadvantage and should not guess the next card but maybe let more cards pass. It is not so clear what happens, but it seems reasonable to adopt a strategy that in some way takes advantage of situations in which fewer red than black cards have passed. Maybe simply to stop me as soon as I have dealt more black than red cards and if this never happens, go for the last card?

Before I give you the answer, let us try this strategy with a reduced deck of two red and two black cards. There are then the following six possible rearrangements:

BBRR, BRBR, BRRB, RBRB, RBBR, RRBB

so the question is in how many of these you win. In the first case, you lose because you will guess "next red" after the first card has passed, but the next card is black. You win in the second and third cases, lose in the fourth (must take the last card), win in the fifth, and lose in the sixth (must take the last card).

Three out of six, winning probability 50%, no improvement over guessing the first card. Try another strategy and guess "next red" after two black cards have passed. With this strategy, you win in the first, second, and fifth cases; in the other cases, you are forced to the last card, which is black. Again, a 50–50 chance. Now try the seemingly stupid strategy to guess "next red" when more red than black cards have passed. Clearly this must be worse? But checking the outcomes reveals that you win in the first and second cases when you are forced to the last card, and you also win in the sixth case when you guess that the second card is red. So the stupid strategy also gives you a 50–50 chance. As a matter of fact, any strategy you choose gives you a 50–50 chance.

And this is also the case with the full deck: You have a 50–50 chance to win *for any strategy*! Any "reasonable strategy" gives an advantage if its deciding condition applies, but this advantage is ruined by the cases in which the condition never shows up. And any "stupid strategy" has advantage and disadvantage switched, but it also has a 50–50 chance. The case of the reduced deck above illustrates this nicely but may not be enough to convince you in the case of the full deck. One argument that is often presented is that rather than guessing "next card red," you might as well guess "last card red." The reason is that whatever your deciding strategy is, at the time it applies, the next card and the last card (which may be the same card) are equally as likely to be red. But the last card is decided already by the shuffle and is equally likely to be red and black. Some people are convinced by this argument, but I have also met those that are not. Of course, there is a strict mathematical proof but this is not the place to present it.

Here is another problem regarding guessing strategies. In each of my hands I have a slip of paper. On each slip I have written a number and I will ask you to choose one hand, look at the number, and guess whether it is the smaller or larger of the two. The numbers are integers (negative or positive), and they are different, but other than that, you do not know anything about what they are or how I have chosen them. Can you do better than a 50–50 chance?

No way, it seems. But surprise, surprise, you can! Here is how to do it. First you decide a "threshold value" by choosing an integer, negative or positive, and adding 0.5 to it. You can choose this integer in any way you want as long as it comes from a wide enough range, essentially spanning all integers, so using a computer is preferable. After you have decided your threshold, you choose one of my hands at random, look at the number, and *assume* that your threshold is *between* this and the other number. Based on this assumption, you make your decision (in essence, you replace my hidden number with your threshold value). If the number you got from my hand is smaller than the threshold, you guess that it is the smaller of the two. If the number you got is larger than the threshold, you guess that it is the larger of the two (and because you have added 0.5, you guarantee that it is not equal to either of the numbers). In this way, the probability that you guess correctly is *larger than* 50%! But why?

There are three cases to consider.

Case I: The threshold is larger than both numbers

Case II: The threshold is smaller than both numbers

Case III: The threshold is between the two numbers

In cases I and II, your assumption that the threshold is between the two numbers is false and you have a 50% chance to guess correctly because the only thing that then matters is what hand you picked. Specifically, in case I, you guess correctly if you manage to pick the smaller of the two numbers; in case II, if you pick the larger. In either case, you have a 50–50 chance. But in case III, when your assumption is actually true, you *always* guess correctly. Taken together then, regardless of how likely you are to be correct in your assumption, your chances are *better than* 50% to guess correctly. Let us treat this a bit more formally. Let G be the event that you guess correctly and A the event that your assumption is correct, and let p be the probability of A. The law of total probability then gives that the probability to guess correctly is

$$\begin{aligned}\mathrm{P(G)} &= \mathrm{P(G\ given\ A)} \times p + \mathrm{P(G\ given\ (not\ A))} \times (1-p)\\ &= 1 \times p + 1/2 \times (1-p) = (1+p)/2\end{aligned}$$

which is greater than $1/2$. One way to test this result empirically is by using a computer or a calculator that has a *random number generator*. A random number generator gives random numbers between 0 and 1, so if you restrict all numbers to be in this range, you can check for yourself. First fix the initial two numbers, for example, 0.3 an 0.7. Then let the computer choose the threshold, which will be a number between 0 and 1. Finally, choose at random between the two numbers 0.3 and 0.7, and apply the guessing rule from above. Repeat this procedure many times and count how many times you guess correctly. The value of p is now the probability that a random number falls between 0.3 and 0.7. As the range of numbers between 0.3 and 0.7 covers 40% of the total range from 0 to 1, we get $p = 0.4$ and the probability to guess correctly is $(1 + 0.4)/2 = 0.7$. It is not difficult to write a computer program that verifies this empirically, but it is still not easy to understand the conclusion of the original problem. You are somehow improving your chances by making an initial guess that is either correct or irrelevant, but other than that, I cannot offer a convincing explanation. Perhaps we should ask Marilyn?

It is fun to present the last two problems together. In the first problem (guess if the next card is red), it seems that you could definitely do better than 50–50 but turns out that you cannot. In the second problem (guess the larger number), it seems impossible to beat 50–50 but turns out that you can. I am sorry to repeat myself, but here I go again: Wasn't that cool?

FINAL WORD

In this section, we have seen example after example that illustrate how our intuition for probability is poorly developed. Perhaps we are somehow conditioned to always look for order and regularity, which makes us overlook and misinterpret many phenomena of chance. Or perhaps we simply don't have enough training? Whatever the reason, we have seen how the laws of probability help us sort out the problems once they have been carefully thought through and properly formulated. We are not done with the peculiarities yet; in the next chapter, we will investigate a particular type of surprising probabilities, the really, really small ones that just won't go away.

3

Tiny Probabilities: Why Are They So Hard to Escape?

PROBABLE IMPROBABILITIES

Chapter 2 dealt with probability problems where the answers may seem surprising and counterintuitive, at least at first. In this chapter, we will continue on this path and focus on a particular oddity: Why do unlikely things happen all the time? Why do people keep winning the lottery and getting struck by lightning when the chances of such events are so minuscule? We have already touched on this in the birthday problem. On page 54 we saw that, in a group of 100 people, we can be almost certain that at least 2 people have the same birthday (the probability is 0.9999997). If we consider two particular individuals, the probability that they share a birthday is $1/365$ (the first can have any birthday; the second must match), which is not very much, only about 0.3%. However, with 100 people, we have $\binom{100}{2} = 4{,}950$ pairs to try for a match, and it is very likely that we finally find one.

The birthday problem illustrates the general idea that although something is highly improbable in the *particular* case, it can still be highly probable in *general* because there are so many particular cases to try out. For another example, consider the state lottery game "Pick 3" where you are asked to match three numbers drawn from 0 to 9, in the order they are drawn or in other words to match one of the thousand combinations ranging from 000 to 999. If you pick three numbers, the probability to win is is $1/1{,}000$, so if you play once, you don't have a very good chance of winning. In some states, for example, Texas, Pick 3 numbers are drawn twice a day, six days a week, so if you play every drawing, you have 624 opportunities to win in a year. This means that you expect to win at least once every two years, so if you keep playing for years, it would be strange if you never won. And when the expected win finally comes, it was still a one-in-a-thousand chance that you would win that particular time.

Let us do some math. Suppose that you repeat some experiment where the probability of success each time is p, and that consecutive successes or failures are independent. If you make n attempts, the probability to have at least one success is

$$\text{P(succeed at least once)} = 1 - (1 - p)^n$$

which is true because $1 - p$ is the probability to fail once, $(1 - p)^n$ the probability to fail n times, and one minus this the probability to succeed at least once, Trick Number One at work again. This is the fundamental formula for this section. Let us try Pick 3 played for a year. Then $p = 0.001$ and $n = 624$, and we get

$$\text{P(win at least once)} = 1 - (1 - 0.001)^{624} \approx 0.46$$

so you have an almost 50–50 chance to win at least once in a year. The times you lose will be quickly forgotten, and the time you win will be the one-in-a-thousand shot that came true.

On a recent *Jeopardy*, it turned out that two of the three contestants had gone to the same high school 20 years apart and host Alex Trebek wondered what the odds of that could be. Again, if we only consider p, the probability that this happens on this particular day, then p is certainly small. However, Alex has hosted Jeopardy for over 20 years so there is a large n here also. And having gone to the same high school is only one type of remarkable coincidence. The contestants could also have gone to the same college, lived on the same street, or been at the same Super Bowl, and in any such situation, Alex would have noticed the coincidence. Let me emphasize that I am only responding to Alex's rhetorical question; I am just as amused as anybody else by fun coincidences.

For extreme odds, consider Evelyn Adams who in the mid-1980s won the New Jersey State Lottery twice within four months, raking in a total of \$5.4 million. The newspapers reported that the chance of this happening was 1 in 17 *trillion* (twelve zeros). But what is this probability, and is it relevant? The probability that Ms. Adams wins twice, buying one ticket each time, can be computed and equals 1 in 17 trillion. But if we instead consider the probability that *somebody* wins *some* state lottery twice in four months, this is much larger. Purdue professors George McCabe and Stephen Samuels took the time to compute this probability and found it to be about $1/30$, not so impressive anymore. And the event that somebody wins some state lottery twice, not necessarily within four months, is more or less certain. Moreover, just like in the case of the Swedish airport manager, we should here really compute the probability of "again" rather than "twice." People win lotteries all the time, and those that win might spend even more money buying tickets afterward (Ms. Adams certainly did). It is actually not unlikely that somebody wins *again*, and there we have it; somebody won *twice.*[1]

An amusing example of how the seemingly extremely unlikely is actually quite likely was given by British physicist Sir James Jeans in his 1940 book *An Introduction to the Kinetic Theory of Gases*. His point was to illustrate the huge number of molecules involved in gas kinetics, and the problem goes as follows: Draw a breath. How likely is it that you just inhaled a molecule that was exhaled by the dying Julius Caesar in his last breath in 44 B.C.?

Seems like there's a zero probability if there ever was one. Let us try to compute it. Sir James teaches us that there are about 10^{44} molecules in the atmosphere and about 10^{22} in a human breath. In the 2,000 years that have passed, we assume that the molecules that Caesar exhaled in his "Et tu Brute" have had time to spread throughout the atmosphere and mix well with the other molecules. Now consider one of your inhaled molecules. What is the probability that it came from Caesar? (You can start breathing now by the way.) This is classical probability. It is simply the number of caesarian molecules in the air divided by the total number of molecules; thus, the probability that you inhale this particular molecule is $10^{22}/10^{44} = 10^{-22}$, as likely as tossing 73 heads in a row. But you have not only one chance to inhale a caesarian molecule but 10^{22} chances, so with $n = 10^{22}$ and $p = 10^{-22}$, the formula above gives

[1]When Maureen Wilcox beat the odds twice, she did not, however, make a dime. In June 1980, she bought tickets for both the Massachusetts Lottery and the Rhode Island Lottery. She managed to pick the winning numbers for both lotteries but didn't win a dime—her Massachusetts numbers won the Rhode Island Lottery, and her Rhode Island numbers won the Massachusetts Lottery.

$$P(\text{inhale at least one caesarian molecule}) = 1 - (1 - 10^{-22})^{10^{22}} \approx 0.63$$

a better than 50–50 chance. The ridiculously small success probability p is compensated for by an equally ridiculously large number of trials n, and in the end, you are more likely to succeed than not. A little warning here. If you try to do the last calculation on a pocket calculator, you will most likely get 0 rather than 0.63. This happens if the calculator first computes what is inside the parentheses and rounds it to 1. A little later I will show you another way to do the computation.

The careful reader might object that the molecules are not independent in the sense that, for example, consecutive coin tosses are. This is true. Indeed, suppose that you inhale the molecules one by one and start by a success. The probability that the next is also a success is then the fraction of caesarian molecules that are left, which has now changed because both the total number and the number of caesarian molecules have decreased by one. This situation is more similar to dealing a poker hand and asking for at least one heart (a 10^{22} card hand dealt from a 10^{44} card deck with 10^{22} hearts) than it is to toss coins and ask for at least one head, and we should instead use combinatorial methods. However, the changes in fractions are so small that we can assume independence for all practical reasons, and this assumption greatly simplifies the calculations.

You may wonder if it is reasonable to assume that all Caesar's exhaled molecules are still around. After all, oxygen is absorbed in the blood stream of people and animals and carbon dioxide is absorbed in plants so these molecules may be removed from the atmosphere for a long time. To be safe we could restrict attention to nitrogen, which makes up 78% of air and is not likely to disappear anywhere. In our calculations, this reduces p by a factor 0.78, and the probability to inhale a caesarian molecule goes down to 54%, still better than 50–50.

Of course, many assumptions about the science of the problem are hard to verify, but, after all, the example is intended to illustrate a point and to entertain and it's not all that important where old JCs exhaled molecules eventually ended up. I bet John Allen Paulos was surprised when after having described the Caesar example in *Innumeracy* was bitterly attacked in a 1993 article in *American Mathematical Monthly*. In this article, a Mr. Renz spends about two-and-a-half pages criticizing Paulos's assumptions and calculations with a tenacity that is exhausting, bordering on the parodic. Can't we all just get along?

SADDAM AND I

In 1987, I went for a swim on the east coast of Australia and lost my glasses in the waves. In 1992, I met Australian statistician Rodney Wolff from the city of Brisbane, not too far from where I went swimming. We started talking about Australia, and after the usual stuff about kangaroos, AC/DC, and vegemite sandwiches, I mentioned my lost glasses. He had not found them.

OK, what kind of a lame story was that? One that is seldom told I can assure you. But what if he had gone for a swim and found my glasses? That would have been an amazing coincidence, and I would have told the story all the time. The point I am trying to make here is that for every amazing story about coincidence that is told, there are countless mundane stories that are never told. Improbable events happen all the time, probable events much more often than that.

Take predictive dreams as an example. You may find it amazing that you dream about Aunt Jane in Pittsburgh only to wake up from her calling you on the phone. It is hard to calculate probabilities of such events, but if she calls you on average once a week and she pops up in your dreams a few times a month, it is not unlikely that the two eventually coincide due to chance alone. And that is the particular event. Now consider all people who have dreams about relatives that then call and wake them up. It probably happens often, at least to somebody, somewhere. When it happens to you, you think it is remarkable.

I found it amusing once to run into an old friend on the Greek island of Patmos. Hardly surprising, though, considering the large number of Swedes who visit the Greek islands in the summer. It is not unlikely that somebody runs into somebody they know. Now it happened to me and that's why I am telling you about it. And once in Sydney, Australia I ran into an old high-school classmate on the street. I found this pretty remarkable, but he wasn't surprised at all; I was the twentieth person he had met from our hometown in the few months he had been there.

There are tons of examples of how the success probability p might be tiny; but if the number of trials n is large enough, it is very likely that there will be a success at last. These examples range from lottery and lightning strikes to mutations, evolution, and life on other planets. Mathematician and popular science writer Amir Aczel has addressed the last question and reports his findings in a book entitled *Probability 1: Why There Must Be Intelligent Life in the Universe.* Such a probability is of course not really possible to compute and you might not agree with his assumptions, but again the point is that a minuscule p is in itself not evidence that something is unlikely to happen. In the case of intelligent life in the universe, it is impossible to fathom how enormous n is. With hundreds of billions of galaxies each with hundreds of billions of stars, we're talking about a huge number of trials here.

Questions about the universe are too profound for me so let me instead tell you the last of my traveling coincidence stories. When I was in college, a friend and I took a year off and traveled to the South Pacific. At the airport in Apia, the capital of Samoa, we heard two people behind us speak Swedish. It turned out that they lived only a few blocks away from my friend. We spent a week together on Samoa and then parted ways. Four months later, we ran into them at the airport in Sydney where they were going on the same flight as us! Probable or improbable I don't know, but it sure was a lot of fun.

When linguists study the relatedness of different languages, they often examine similarities between words that are believed to be old in the language, for example, words for numbers, family relations, and body parts. Sometimes such similarities show up by coincidence. For example, the word in modern Greek for "eye" is "mati" (μάτι) and the Samoan word is "mata." Although this might seem like an unlikely coincidence, I have yet to hear a linguist suggest that the languages are related.[2] Nor have I heard anybody claim a link between the ancient Greeks and the Chumash Indians of California although both named places "Simi" (Greek island Σύμη and the wine-producing Simi Valley). It is simply not that unlikely to find *some* words that are similar in *some* unrelated languages. There is only a limited number of two syllable words and plenty of languages to try so eventually there will likely be a match. Let me conclude the Polynesian language lesson by mentioning that the Samoan word for "four" is "fa" just like the fourth "sol-fa syllable" (do-re-mi-fa), and you can use this to concoct your own linguistic-musicological conspiracy theory. Finally, my favorite linguistic coincidence: The river Potomac was named by Native Americans and the Greek word "potamaki" (ποταμάκι) means "little river." What are the odds?

A collection of remarkable coincidences in daily life (including the stories of Evelyn Adams and poor Maureen Wilcox from the previous section) is presented in the book *Beyond Coincidence* by Martin Plimmer and Brian King. It is not a mathematical book, and they do not present much probability, but

[2]Even after they find out that there is a Greek island named Samos (home of ancient math superstar Pythagoras).

some of their examples can be analyzed with methods we have learned. For example, they report the eerie fact that the winning Pick 3 numbers in the New York State Lottery on September 11, 2002, were...yes, 9–1–1. Numerological significance or just probabilistic mundanity? Remarkable in any case. Plimmer and King also teach us that Georg Friedrich Händel and Jimi Hendrix lived next door to each other in London, over 200 years apart. One wonders if they would have gotten along. I'd like to think so.

The lottery game *Cash Five* can be played in several American states. You choose five numbers from, for example, the numbers 1 through 40 (this differs between states). Would you rather choose 1–2–3–4–5 or 3–11–14–26–39? Many would prefer the second. Come on now, the first five numbers in a row? That will never happen! But each particular choice of five numbers has the same probability. Remember the combinatorics section to realize that there are $\binom{40}{5} = 658{,}008$ ways to choose the five numbers, and thus each particular choice of five numbers has probability $1/658{,}008$. It is just as unlikely to get the first as the second sequence of numbers. Still, I think many would instinctively say that 1–2–3–4–5 is less likely than 3–11–14–26–39. The first sequence has a very nice and recognizable feature; it is the first five numbers. The second, on the other hand, does not display any particular pattern and we do not think of it any differently than we would of, for example, the sequence 5–8–19–24–33. The sequence 3–11–14–26–39 for us just represents any sequence where there is no particular pattern, and getting *some* such sequence is of course much more likely than getting the first five numbers in a row. Any sequence of five numbers in a row will undoubtedly be noticed and considered "remarkable." There are other remarkable sequences, for example, 1–3–5–7–9 or 2–4–6–8–10, but there are far more unremarkable sequences and that is why we tend to believe that an individual representative of these is more likely than an individual representative of the remarkable sequences.

The same is true for coin tosses. If you toss a coin ten times in a row, the sequences H H H H H H H H H H and H H T H T T T H H T are equally likely with probability $(1/2)^{10}$ each, but only the first would be considered remarkable. We tend to think of the second as some random ordering of five heads and five tails, and there is nothing remarkable about getting five heads and five tails (it has probability ≈ 0.25). However, the probability to get them in *exactly this order* is still $(1/2)^{10}$, the same as ten heads in a row.

It seems to be a deeply rooted instinct in us humans to look for patterns, regularities, and coincidences and some take it far. Secret messages and conspiracies are being conveyed through numbers or letter combinations. During our Fall 2005 hurricane evacuation tour, my wife and I spent some time in and around El Paso, Texas (we figured that the day a hurricane struck there would be the day of Armageddon anyway). One day I was sitting at a table on the patio of the excellent restaurant Ardovino's Desert Crossing, reading Peter Bernstein's *Against the Gods*, a splendid account of the history of probability and statistics from the perspective of risk management. I was enjoying a glass of Roswell Alien Amber Ale in the West Texas winter sun when I suddenly realized that "Bernstein" is the German word for "amber." Spooky! I'm still trying to figure out what those Roswell aliens were trying to tell me.

Another typical example of coincidence is the common acquaintance. I am sure it has happened to you; you start talking to somebody on an airplane and find out that his cousin is your sister's hairdresser. Or something. I once met a Swedish guy on the island of Tonga, also named Peter by the way. It turned out that he had lived in my hometown as a child and that we had a common childhood friend, Magnus. This is an instance of the *small-world phenomenon*, the hypothesis that any two people in the world can be connected through a short link of acquaintances. A pioneering experiment was done in 1967 by psychologist Stanley Milgram. He sent a letter to a group of volunteers in the Midwest and asked them to make sure that it reached a specified person in Massachusetts. However, they were only allowed to deliver the letter to a personal acquaintance and the task then passed on to this person and so on until it reached the target individual. The subjects thus had to try to figure out who of their friends that would be most likely to be able to reach the target of whom they only knew name, location, and occupation. Only 15% of the letters eventually reached the target individual, but among those that did, no chain was longer than 11 steps and the average length was 8. The experiment has since been repeated several times both by Milgram and others, with regular mail and email, and a typical number of six steps has stuck in the folklore. You may have heard the expression "six degrees of separation," and it is sometimes even heard that the small-world phenomenon means that everybody can be connected to everybody else in exactly six steps. This is nonsense. There is nothing scientific about the number six, but the point is that many people can be connected with each other by chains that tend to be surprisingly short.

In smaller and more specialized communities, degrees of separation can be computed with more accuracy. In footnote 3 on page 51 in Chapter 2, I mentioned mathematician Paul Erdős. Erdős was one of the most prolific mathematicians ever with about 1,500 papers published in scientific journals and about 500 co-authors, and this has inspired the concept of *Erdős number*, the number of steps needed to link a mathematician to Erdős through research publications. Erdős himself has number 0. Somebody who has co-authored a paper with Erdős has Erdős number 1; somebody who has co-authored a paper with somebody who has co-authored a paper with Erdős has Erdős number 2; and so on. Most mathematicians have an Erdős number, and among those, the average is about 4.7. One reason that Erdős had so many co-authors is his eccentric lifestyle. He did not have a permanent address but lived out of a suitcase (or two), traveled around the world, and stayed with mathematicians who would provide him with room, board, and coffee in exchange for collaborative mathematical research.

My own Erdős number is 4. In 2001, I wrote a paper about branching processes (remember the discussion on page 66) applied to a dataset of centenarians (hundred-year-olds) in Massachusetts together with, among others, renowned statistical geneticist Ranajit Chakraborty. He has written a paper with Indian statistician C. R. Rao (a very famous man in statistics circles) who in turn has written a paper with astrostatistician (yes!) Jogesh Babu who has published a paper with Erdős. There is nothing special about having an Erdős number 4; over 40,000 people have number 3 or lower. There is a large online database of mathematical publications called *MathSciNet* that has, other than its more serious search criteria, a feature that gives you the Erdős number for any mathematician of your choice.

In Hollywood, Erdős numbers are replaced by Kevin Bacon numbers and a link between two actors is established when they appear in the same movie. The average Bacon number is 2.95. It is lower than the average Erdős number because there are many more actors in a movie than there are co-authors on a math paper. Chicago probabilist Patrick Billingsley has Erdős number 4 but is also member of an exclusive club, namely those who have both Erdős and Bacon numbers. An appearance in *The Untouchables* gave Mr. Billingsley a Bacon number of two (linked to Kevin Bacon via, for example, Robert De Niro and the movie *Sleepers*). According to the *The Oracle of Bacon at Virginia* website, there are over 1,000 actors who would make a better "center of the Hollywood universe" than Kevin Bacon, for example, Michelle Pfeiffer (the average "Pfeiffer number" is 2.88), Diane Keaton (2.82), and Clint Eastwood (2.80). All famous actors are very close, more or less tied, and I suppose that the tie-breaker was the catchy expression "six degrees of Kevin Bacon." And although a fine actor, Mr. Bacon is no Hollywood analog of Paul Erdős. For that we would need to combine James Dean, John Wayne, and Marlon Brando to get the right blend of talent, productivity, and eccentricity.

What is your Saddam number? If a connection is defined by a handshake, mine is 4. The crucial intermediary is former Secretary of State James Baker III who spent a lot of time on the Rice University campus when I worked there. I am sure that most of my cousins have no idea that they have Saddam number 5. My Saddam number may well be lower though, and there is to my knowledge no equivalent of the MathSciNet database for former dictators.

By now you are probably under the impression that nothing can surprise a probabilist, and that I am trying to convince you that nothing should surprise you either. Whether you are struck by lightning or win the lottery, should you just make some quick calculations and stoically notice that it was bound to happen? I will not tell you how to live your life, but you may test yourself on the following little quiz. You are walking down the street and suddenly run into this very attractive (wo)man whom you had a crush on in high school. No wedding ring, big smile, (s)he hugs you and exclaims something about what an incredible coincidence it is to run into you after all these years. Choose the correct reply: (a) "Yes, that is amazing! Let's go for a drink!" or (b) "Well, statistically speaking..."

TAKING TINY RISKS

Most of the tiny probabilities we have looked at so far can be computed in a straightforward manner. Other tiny probabilities must be estimated from data (remember how I mentioned statistical probabilities on page 3). This is, for example, the case when various risks are calculated (and insurance companies do such calculations all the time). In his 1992 book *Chances: Risk and Odds in Everyday Life*, Mr. James Burke teaches us that the probability that you are struck by lightning is 1 in 600,000 and the probability that you are involved in a fatal plane crash (but not necessarily get killed yourself) is 1.6 in 10 million. The only way to arrive at such figures is to use statistical methods (which you will learn more about in Chapter 8). The lightning probability is simply the number of people struck by lightning in some year in some country (in Mr. Burke's case, I presume the United Kingdom) divided by the total population. Of course, the true risk for you depends on who you are and what you do. If you are a golfer and don't shy bad weather, you are in a high-risk group.[3] If you never go outside in a thunderstorm, you are pretty safe. And if you should be struck, you still have a good chance to survive because only about 10% of lightning strikes are fatal. The plane crash probability is computed in a similar way as the number of flights that have experienced a fatality divided by the total number of flights.

When tiny probabilities are based on data, they can fluctuate wildly, especially if the data are scarce. In aviation safety, the *fatal event rate* is defined as the number of fatal events per million flights (because start and landing are by far the riskiest moments during a flight, this makes more sense than relating fatal events to the number of miles traveled). The Concorde was once the world's safest aircraft with no fatal accidents at all, and then came the crash on July 25, 2000. As there were so few Concorde flights, about 80,000, the fatal event rate went up from 0 to $1/80{,}000$ or about 12 per million, the worst safety record of any aircraft model. In an instant, the Concorde went from being the safest to the most unsafe aircraft! Compare its figures with those of the most traveled aircraft, the Boeing 737. This model currently has about 105 million flights and a fatal event rate of 0.41 per million. Another fatal crash with a 737 will not change the rate by much. Because of this stability, the figures for the Boeing 737 are more meaningful than those for the Concorde.

The airplane example illustrates how it is hard to understand and adequately compare tiny probabilities. Another example: In 2000, the murder rate in Sweden was twice that of Switzerland. Sweden had 1.97 murders per 100,000 people and Switzerland 0.96 (for comparison, the United States had a murder rate of 5.64 and South Africa topped the list with 50.14). It may be "statistically true" that you are twice as likely to be murdered in Sweden as in Switzerland, but it should hardly affect your vacation plans. In *Chances*, James Burke reports that Luxembourg in 1992 had the highest murder rate in Europe. In 2000, Luxembourg was low on the list. Such fluctuations are typical for a country with a small population; a small change in the *number* of murders can lead to a big change in the *rate*. Just look at this example: In 1998, Swiss Guard commander Alois Estermann and his wife were murdered in Vatican City, making its murder rate go up from its usual 0 to over 200 per 100,000, dwarfing the rest of the world. The next year it was down to 0 again. But even

[3]By the way, lightning does strike twice and in the case of Roy Sullivan, seven times! This U.S. park ranger had a particularly complicated relation to Thor, but the God of Thunder did not kill him. In 1983, Sullivan took his own life for reasons unrelated to meteorology.

when populations are large and estimates more reliable, tiny risks remain tiny even if they double or triple.

What is the risk of dying from being hit by a meteorite? Hard to say because it has not happened to anybody yet. A dog was killed by a meteorite in Egypt in 1911 and Ann Hodges of Sylacauga, Alabama was hit by one in 1954 but survived. These are extremely scarce data, and an actuary would have problems estimating this risk for your life insurance policy. People have died from other falling objects though. In 456 B.C., an eagle flew over ancient Greek playwright Aeschylus and dropped a tortoise on his head, killing him on the spot. Don't worry about meteorites. Statistically, you are more likely to be killed by a falling tortoise.

A MILLION-TO-ONE SHOT, DOC, MILLION TO ONE!

When it comes to describing unlikely events in daily language, the phrase "one in a million" is the absolute favorite. A quick Internet search reveals that it is about 20 times more common than "one in a trillion," "one in a billion," "one in a hundred thousand," and "one in ten thousand" combined. And search for "one in a zillion," and you are asked if you didn't really mean million. According to *Seinfeld's* Kramer, the phrase "It was a million-to-one shot, doc, million to one!" ends every "proctologist story" (for details you have to check out the episode "Fusilli Jerry" for yourself).

So why a million? There is, of course, the language aspect: "Million" rolls off the tongue easier than the other numbers. I also suppose that a million can be considered a large number but still one that we are comfortable with. Familiar examples where numbers are in the millions include populations of nations and large cities, budgets for medium-sized cities, box office revenues for popular movies, and Tiger Woods' salary. I have found some evidence of "number inflation." In 1869, our old friend Sir Francis Galton wrote the following about his definition of an "eminent" man:

> When I speak of an eminent man, I mean one who has achieved a position that is attained by only 250 persons in each million of men, or by one person in each 4,000. 4,000 is a very large number—difficult for persons to realize who are not accustomed to deal with great assemblages.
>
> Sir Francis Galton, *Hereditary Genius*, 1869

Those who qualified as one in a million, Galton labeled "illustrious," adding that a million is "a number so enormous as to be difficult to conceive." Maybe as time goes by, and inflation with it, we will get more and more accustomed to billions and Kramer will be voiced over in future re-releases of the Seinfeld DVDs.

So how much is one in a million? The fast-quipping Frank Drebin (Leslie Nielsen) from the *Naked Gun* movies makes the following observation:

> Jane: Your chances are one in a million.
> Frank Drebin: Better than any state lottery.
>
> *The Naked Gun 33 1/3: The Final Insult*, 1994

Certainly true, at least for jackpots in the high-prize games. The really big money is in the multi-state lotteries *Powerball* (currently 27 states) and *MEGA Millions* (currently 12 states). In Powerball, you choose five numbers from 1 to 55 and one "power ball number" from 1 to 42 so there are

$$\binom{55}{5} \times 42 = 146{,}107{,}962$$

possible combinations. If you want the one-in-a-million chance to win, you must play 146 different combinations. In MEGA Millions, the odds are even higher. You pick five numbers from 1 to 56 and one number from 1 to 46, which gives

$$\binom{56}{5} \times 46 = 175{,}711{,}536$$

different combinations, so you have to pay \$176 to get your million-to-one shot. Of course, since there are many minor prizes for partial matches, the chance to win *something* is much higher, about $1/40$.

If you play bridge, you have about a one-in-a-million shot to be dealt ten hearts and one card each in the other suits. In poker, you have a better than a one-in-a-million shot to be dealt a royal flush (10–ace in the same suit), about 1.5 in a million. And when Sean Connery saw his number—17—come up three times in a row in a casino in Italy in 1963, this was far more likely, a 20-in-a-million shot.

In some regard, one in a million is not that impressive. If we talk about people and say that something happens to just one in a million, it still happens to well over 1,000 Chinese, almost 300 Americans, 9 Swedes, and even almost to some dude from Fiji.

The acclaimed British twentieth-century mathematician J. E. Littlewood defined a *miracle* as "an event that has special significance when it occurs, but occurs with a probability of one in a million." He then went on to calculate that people experience miracles at a rate of roughly one per month, which has become known as "Littlewood's law of miracles." You can do even better. If you flip a coin 20 times, you will get a sequence of heads and tails. I just did it and got the seemingly unremarkable sequence

HTTHHTHTTHTTTHHTHTTT

However, each particular sequence of heads and tails has probability $(1/2)^{20}$, which is about one in a million. My sequence had 8 heads and 12 tails, which is not that unlikely, but the chance of getting them in exactly this order is still one in a million and I think it bears special significance that I managed to achieve something so highly improbable. Start your morning with a cup of coffee and 20 coin flips, and you can make every day miraculous.

We have gotten a pretty good grip of how much is one in a million. I will leave it as an exercise for you to compute a snowball's chance in hell.

MONSIEUR POISSON AND THE MYSTERIOUS NUMBER 37

If you randomly place 64 grains of rice on a chessboard, what proportion of squares can you expect to be empty? If you have a one-in-a-million chance to succeed with something and try a million times, what is the probability that you still do not succeed? If you select a square mile at random in Jackson County, Colorado and walk around, what is the probability that you don't run into anybody? And if you are told that large meteorites hit earth on average once every 1,000 years, what is the probability that there will not be any meteorite hits within the next millennium?

And the answer to all those questions is 37%. We also saw this number show up in the online shopping example on page 62 where your chance of winning is about 37% with the optimal strategy of letting the first 37% of the offers pass and take the next that is better. And in the Caesar example on page 74, the chance of *avoiding* caesarian molecules is 37%. What then is so special about the number 37? My friend Leif, a retired Swedish number theorist, has often patiently pointed out to me that 37 is the smallest irregular prime number and, at the same time, the largest idoneal prime number. I have no idea what this means. Thirty-seven is also the normal human body temperature in degrees Celsius and the ideal age for a man to get married, at least according to Aristotle. But number theory and Greek philosophers aside, the number 37 really shows up because 0.37 is the approximative value of the number e^{-1}. If you are not familiar with this last mathematical expression, let me explain quickly. The number e is the base of the so-called *natural logarithm* and is arguably the most important number in mathematics (although π might object), indispensable to mathematicians. The numerical value of e is 2.718281828459... where the decimals continue forever with no particular pattern.[4] The notation e^{-1} is a mathematician's way of writing $1/e$, and this equals 0.3679... or approximately 0.37.

OK, so that was not so mysterious, but why does the number e^{-1} show up in all of these examples? It has to do with something that is called the *law of rare events*. This law says that if an event is rare, unpredictable, and occurs on average once, then the probability that it does not occur at all is e^{-1}. This applies to all of the above examples. Fix 1 of the 64 squares on the chessboard, for example, the square **a1** (the corner square closest to the white player's left hand). The probability to hit **a1** with a grain of rice is $1/64$, which is fairly small. It is totally unpredictable when a hit will come; it can occur anytime, independently of how many times you have hit **a1** before. As you try 64 times, you expect to hit once, but that is about all you can say for certain. Of course, you may still fail to hit and the law of rare events tells you that the probability of this is e^{-1}. This also means that we can expect this proportion of squares, about 37%, to be void of rice grains. Similarly with your million-in-one chance: rare, unpredictable, and occurs on average once in a million attempts. The population density of Jackson County, Colorado is one per square mile, so if you choose a location at random, bumping into somebody can be considered a rare event. However, if you walk your entire chosen square mile, you expect to run into on average one person. As people are not uniformly spread and it is (somewhat) unpredictable where they are, the law of rare events applies again and the probability of walking an empty square mile is 37%. A meteorite hit certainly qualifies as rare and unpredictable, and if it happens once in a millennium, the probability that any particular millennium is spared is 37%. And, come to think of it, 37% of us will never experience that "once-in-a-lifetime" opportunity.

What if the events occur on average twice, or three times, or some other number of times? And what if we instead want to ask for the probability that

[4]When Google went public in 2004, the company famously said in their SEC filing that they expected to raise $2.718281828 billion from the initial public offering of shares. They fell short of their goal by about a billion but got high style points from mathematicians.

it happens once, or twice, or some other number of times? Actually, the law of rare events is more general than I said above, and it does cover these cases as well. Thus, suppose that we are dealing with some rare, unpredictable event that occurs on average λ (Greek letter "lambda") times. The number of occurrences is said to follow the *Poisson distribution*, and the probability of k occurrences for $k = 0, 1, 2, ...$ is given by the formula

$$P(k \text{ occurrences}) = e^{-\lambda} \times \frac{\lambda^k}{k!}$$

It is quite surprising that the vague assumptions of rare and unpredictable events can give such a precise formula for the probability. To deduce the formula, certain mathematical assumptions must be made that formalize these verbal assumptions, but this is not the place to do it. The number λ is called the parameter of the Poisson distribution (recall how we also used this terminology for the binomial distribution on page 39). In the examples above, I arranged it so that λ was always equal to 1; hence, the probability of 0 occurrences equals e^{-1} (for the second factor in the probability above, remember that anything raised to the power 0 equals one, and that 0! also equals one). If we instead randomly place 128 grains of rice on the chessboard, we have on average two grains per square, which gives $\lambda = 2$ and as $e^{-2} \approx 0.14$, we can expect about 14% of the squares to be empty. If we count the number of meteorite hits during a century instead of a millennium, we get $\lambda = 0.1$ and the probability of no hits equal to $e^{-0.1} \approx 0.9$ and so on.

The Poisson distribution fits many real-world datasets with good precision. Examples that are commonly given are the number of misprints on a page in a book or newspaper, the number of mutations in some section of a DNA molecule, counts of radioactive decay, the number of stars in some large volume of space, and the number of hits of some specified webpage. Wait a minute now, hits of a webpage? Yahoo, Google, Amazon? Not exactly rare are they? Well, it depends on the time scale. You can always find a time scale that makes it rare, milliseconds or microseconds, or whatever you want. As hits are unpredictable, coming from a large number of users acting independently of each other and spread all over the world, the Poisson distribution still fits well. A general observation is that "rare" is not enough without "unpredictable." Once you have determined the proper time scale to make your events rare, they cannot exhibit any kinds of regularities. Note the obvious difference between the statements "large meteorites hit earth once every thousand years" and "a millennium year occurs once every thousand years." Both are rare; only the first is unpredictable.

The Poisson distribution is named for French mathematician, Siméon-Denis Poisson (1781–1840), who published his results about the distribution in 1838. One of the early applications of the Poisson distribution was to the number of Prussian soldiers annually kicked to death by their horses, by now a classic example. The fit to a Poisson distribution is remarkably good (and gives our use of the term "success" an interesting meaning). This is not quite like placing grains on a chessboard where the success probability is the same all the time. Some horses may be more aggressive or more skillful in their kicking, but the Poisson distribution works fine anyway as long as the probabilities of success do not vary too dramatically.

In the example with the rice grains on the chessboard, we can of course also compute the probability that **a1** is rice-free with the formula from page 72. This formula with $p = 1/64$ and $n = 64$ gives

$$\text{P(square } \mathbf{a1} \text{ is free of rice)} = (1 - 1/64)^{64} \approx 0.37$$

which is approximately equal to e^{-1}. In the same way, you have probability e^{-1} not to succeed if you repeat your "one-in-a-million shot" a million times; just replace 64 by one million in the formula. It is a well-known mathematical fact that

$$(1 - 1/n)^n \approx e^{-1}$$

if n is large. A more precise mathematical formulation is that e^{-1} is the *limit* of $(1 - 1/n)^n$ as n *goes to infinity*, which means that the "$\approx$" in the formula becomes more and more precise the larger n gets. Try it for yourself with different values of n if you wish. This result offers an alternative way of computing the probability in the Caesar example above. With $n = 10^{22}$, we get the probability to avoid caesarian molecules as e^{-1}. In this example, it is a coincidence that p equals $1/n$, and I suppose that Sir James arranged it so that the expected number of inhaled caesarian molecules would be one. This allowed him to claim that "chances are that each of us inhales one molecule of it with every breath we take" (Sir James did not actually do the calculations I have done).

The attentive reader may have noticed that we are now talking about the Poisson distribution in some cases that we could have treated with the binomial distribution. This observation is correct, and there is a general result that says that if you have a binomial distribution with a small p and a large n, then you can instead use a Poisson distribution with $\lambda = n \times p$. This gives an approximation that has the advantage of simpler calculations because you avoid computing potentially huge binomial coefficients $\binom{n}{k}$ and avoid problems with rounding that I mentioned in the Caesar example.

Let us finish with an example of *matching*. Fix an integer n, and write down the numbers 1, 2,..., n in random order. Call it a "match" if the number k is in the kth position. For example, with $n = 5$, there are two matches in the sequence 32541 (2 and 4) and none in 23451. What is the probability that there are no matches at all?

First of all, this probability obviously depends on n so let us try some simple cases. If $n = 2$, we get two matches for the sequence 12 and none for the sequence 21. Probability of no matches: $1/2$. Next, take $n = 3$. There are six ways to list the numbers 1, 2, 3 in order: 123, 132, 213, 231, 312, 321, and in two of these (231 and 312), there are no matches. Probability $1/3$. With four numbers, there are 24 combinations, of which it is readily checked that 9 have no matches, so the probability is $9/24 = 0.375$. With five numbers, the probability of no matches is $44/120 \approx 0.37$, and after that, it stays very close to 0.37 forever. Our old friend 37% shows up again, perhaps not so mysteriously anymore. If we consider an arbitrary position, the probability of getting a match is $1/n$ (convince yourself about this), and if n is fairly large, such a match is a rare event. As we are trying n times to get a match, we can expect this to happen on average once, and it looks like the law of rare events applies here also. The difference between this problem and the examples above is that successive matches are not independent. For example, if there is a match in the first position, the conditional probability of a match in the second position is $1/(n-1)$ and not $1/n$. It can be proved mathematically that the law of rare events applies provided the dependence is not too strong, like in this case. If we consider the *number* of matches, this follows a Poisson distribution, at least for a number of matches that is small relative to n. Obviously, with n numbers, it is impossible to have $n - 1$ matches, so the Poisson distribution does not work perfectly.

The matching problem is a classic in the probability literature and is often stated in terms of the clumsy secretary who puts letters at random in addressed envelopes, or in terms of a group of men who leave their hats at a party and pick them at random when they leave. The probability that at least one letter ends up in the correct envelope and the probability that at least one man gets his own hat are both 63%, regardless of the number of letters or men. Quite interesting and not a result that comes right off the top of the head.

The number 37 has been demystified. It is, however, still undoubtedly the largest idoneal prime number.

CLUMPS IN SPACE

One particular feature about rare and unpredictable events is that they often tend to occur in clumps. This may be counterintuitive to many; if a rare event has just occurred, we don't expect it to happen again for a long time. And conversely, if it has been a long time since the last occurrence, we expect it to happen soon. When we lived in Houston, I often heard people say that the city was "due for a big hurricane." After Houston, we moved to New Orleans, three weeks before Hurricane Katrina. The Orleanians were more relaxed, preferring to point out that the city had not been seriously hit in a long time and concentrating instead on the hurricanes served in tall glasses at Pat O'Briens. I do appreciate their attitude but not the quality of the levee system. Anyway, the point here is that hurricanes (rare on a short time scale and unpredictable at least until they start organizing) hit regardless of where and how hard they have hit before. You are not "due" if it hasn't happened, and you are not safe if it has (just ask the people of Florida). It happens when it happens, period.

The same is true for a lot of things. Suppose that your doctor told you that you have a serious disease and gives you a 50–50 chance to survive. "However," he goes on, "you are lucky since my last patient died." You'd be out of there quickly. As another illustrative example, let us take roulette. Each number comes up on average once every 38 spins, but of course you don't expect *exactly* 38 spins between each occurrence. What is the typical pattern for occurrences of an individual number? I did a computer simulation of hundreds of roulette spins and kept track of the number of spins between successive occurrences of the number 29. Here are the lengths of the first ten such periods:

8 43 20 77 52 6 9 162 22 30

Notice the wild fluctuations. At one point it took 162 spins to get 29 again; at another, only 6. The average of the numbers above is about 43, close enough to 38 to indicate that this is not necessarily anything unusual. In fact, it is when

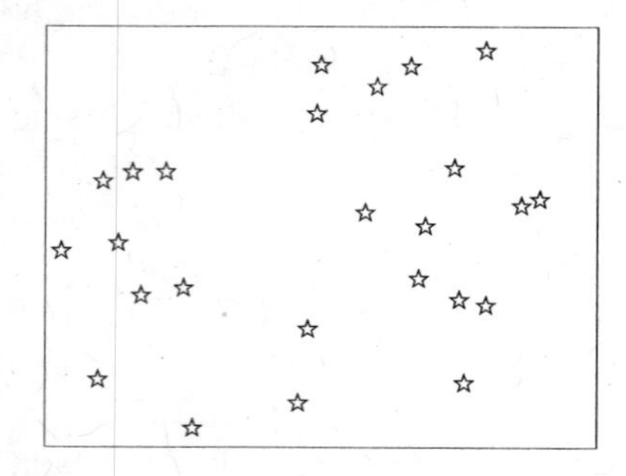

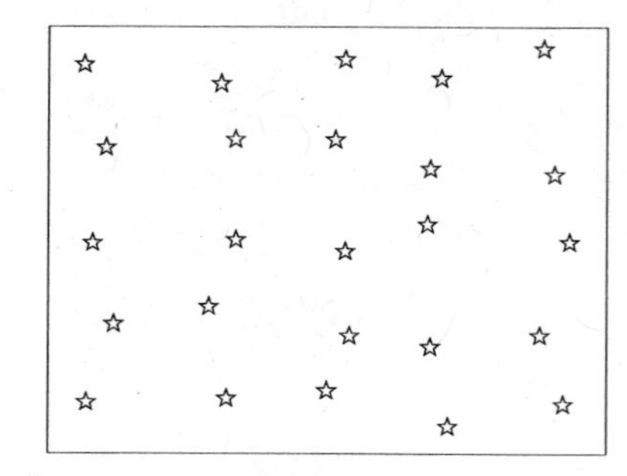

Figure 3.1 Two simulated starry skies. The stars to the left are randomly placed, and the stars to the right are placed systematically with small random variations added.

you observe something that looks too regular that you should be suspicious. Look at this sequence, for example:

35 29 28 44 46 40 45 50 25 47

which has an average of 39, also close to 38. However, this sequence is not likely to have been generated by a roulette wheel, real or computer simulated because the deviations from 38 are too small and there is far too much regularity. The sequence was constructed by me in an arbitrary but not at all random way.

The stars in the night sky provide a nice and easily observed example of clumping. If you look up, stars tend to be clumped and clustered with areas of empty space inbetween. Of course, because of gravitation, the stars do not act totally independently of one another but the dependence is weak enough to give an approximate Poisson distribution of the stars in space. It would certainly be surprising if one day Orion (or the Southern Cross if you are a reader in the Southern Hemisphere) was gone and the stars were neatly organized on a grid. In Figure 3.1 there are two computer-simulated starry skies. The 25 stars to the left are placed completely at random, whereas the stars to the right are placed on a grid with some small random variations. I think you agree that the random picture much more resembles a piece of the real night sky. Randomness does *not* create regularity, as a matter of fact quite the opposite. No wonder that ancient peoples saw patterns in the sky and named them for various animals or objects. If you look at my random stars, you can probably make up several constellations of your own.

For an example of something that is neither completely random nor entirely regular, consider buses that arrive at a bus stop. There is a certain amount of randomness in their arrival times, but as they attempt to run on a schedule, there is far too much regularity to observe a Poisson distribution of the number of buses in a given time interval. Arrivals may be rare on some appropriate time scale but they are not unpredictable enough for the law of rare events to work.

FINAL WORD

This chapter has dealt with perhaps the most fascinating types of probabilities: the really tiny ones. The fundamental lesson to be learned is that they just can't be avoided. Just because an event is rare does not mean that it will not happen; as a matter of fact, extremely rare events occur all the time. In conjunction with noting that a probability is tiny, you must also ask yourself how many times the experiment it describes has been tried. Regardless of how small the chances are, if you keep trying, eventually it will happen. Just ask Evelyn Adams. Or Julius Caesar.

4

Backward Probabilities: The Reverend Bayes to Our Rescue

DRIVING MISS DAISY

In this chapter, we will talk more about conditional probabilities. Recall from earlier that we can think of a conditional probability as the probability of an event if we are given some additional information. So that you do not have to flip back, let me state the fundamental equation for conditional probabilities once more. The conditional probability of the event B, given the event A satisfies

$$P(\text{B given A}) = \frac{P(\text{A and B})}{P(\text{A})}$$

and recall that this is different from P(B) only if A and B are dependent events. If A and B are independent, they have no impact on each other's probabilities. In other words, to condition on an independent event is to incorporate irrelevant information.

We can of course also compute the conditional probability in the other direction, the conditional probability of A given B. As "B and A" is the same as "A and B" and P(A and B) can be computed as P(B given A) × P(A), we get the following expression:

$$P(\text{A given B}) = \frac{P(\text{B given A}) \times P(\text{A})}{P(\text{B})}$$

a formula that is known by the name *Bayes' rule* (or Bayes' theorem), named after the Reverend Thomas Bayes (1702–1761).. Bayes himself did not publish his rule, and it was only published posthumously after a friend had found it among Bayes' papers after his death. This little innocent-looking formula by an eighteenth-century English clergyman and lay mathematician has had enormous influence on scientific and statistical thinking. The reason for its usefulness is that it may be easy to compute the conditional probability in one direction, P(B given A), but what you really want is to compute it in the other direction, P(A given B). Bayes' rule gives a recipe for how to do the calculation. There are more complicated versions of Bayes' rule, and its modern applications include email spam filters and Internet search engines. We will mostly focus on how Bayes' rule is invaluable when it comes to evaluating evidence, for example, in medical tests and in the courts. But first, let us return to the neighborhood pub.

You offer to drive Daisy the drunk home, but she objects. She has just heard on the news that drunk drivers are responsible for 25% of car accidents. Daisy wants to drive. "You sober people cause 75% of the accidents," she slurs, "you should stay off the roads." You suggest that you call Stevie Wonder and let him drive. After all, no accidents involving blind musicians driving have been reported in your city in a long time.

Your Stevie Wonder comment amuses Daisy enough that she lets you drive, but you are still not sure now to argue against her statement. Bayes' rule will set things straight, but we need some numbers. Let us make some up and suppose that, on the streets around your pub, 95% of all drivers are sober and that the overall risk of having an accident is 1%. The relevant probabilities for you and Daisy are then the probabilities that you have an accident if you are sober, P(A given S), and the conditional probability to have an accident if you are drunk, P(A given D). The probabilities we are given are the reversed: $P(\text{D given A}) = 0.25$ and consequently $P(\text{S given A}) = 0.75$. We also have $P(D) = 0.05$, $P(S) = 0.95$, and $P(A) = 0.01$. Let us start with your accident probability P(A given S):

$$P(\text{A given S}) = \frac{P(\text{S given A}) \times P(A)}{P(S)} = \frac{0.75 \times 0.01}{0.95} \approx 0.008$$

so you have a 0.8% risk of having an accident. On the other hand, Daisy has probability

$$P(\text{A given D}) = \frac{P(\text{D given A}) \times P(\text{A})}{P(\text{D})} = \frac{0.25 \times 0.01}{0.05} = 0.05$$

and with her 5% accident risk, she should of course be in the passenger seat (or back seat preferably). She confuses the probabilities P(D given A) and P(A given D), a common mistake. That drunk drivers are responsible for 25% of the accidents is not a relevant figure until we also know how many drunk drivers there are. As they make up much less than 25% of the drivers, they are involved in disproportionally many accidents.

Note that the probability that you are interested in, P(A given D), would not be possible to compute directly from existing data based on accident records. If you know how many accidents and how many drivers there are, you can figure out how likely it is overall to have an accident. From police records on breath and blood tests, you can figure out how likely it is that somebody in an accident is drunk. There is, however, no way to directly compute the probability that a drunk driver has an accident because you have no record of those who don't get into accidents. Bayes' rule helps you do precisely that, if you have some idea about the proportion of drunk drivers (which may be learned from police checkpoints where such are permitted or from surveys where people are asked anonymously about their drinking and driving habits).

Confusion of a conditional probability in one direction with the conditional probability in the other direction is a common phenomenon. When I was a young man, I liked to read Ed McBain's detective stories about the eighty-seventh precinct in the fictional city of Isola. I remember one episode of an Italian-American who was disappointed that almost all Italians on a particular TV show were portrayed as crooks. He wrote to the TV station and voiced his complaints and got a reply that said something like, "That is not true, we also have many criminals who are Jewish and Irish." It was pointed out in the book that the letter-writer was clever enough to distinguish between the phrases "All Italians are crooks" and "All crooks are Italian." This is not a probability problem, but we can think of it as one. On the TV show, the relevant conditional probability P(crook given Italian) is close to one, whereas the irrelevant conditional probability P(Italian given crook) may be much smaller.

I recently read the following question regarding AIDS: "What is the risk of contracting AIDS from a blood transfusion?" and the answer that followed was that 2% of AIDS cases are thought to be caused by blood transfusions. Notice how the question is about the conditional probability P(AIDS given blood transfusion), but the answer is about the reversed conditional probability,

P(blood transfusion given AIDS). Of course, a 2% risk to contract HIV would be far too high for anybody to safely have a blood transfusion, and the relevant probability is of course much smaller, reportedly much less than one in a million. The 2% figure only means that among those who have AIDS, 98% got it in some other way. This information is irrelevant for anybody who must decide whether to accept a blood transfusion.

And to all bachelors and bachelorettes out there: Don't get married. One hundred percent of all divorces start with marriage.

BAYES, BALLS, AND BOYS (AND GIRLS)

Before we get to serious applications in medicine and law, let us consider the following simple problem. An urn contains two white and two black balls. You draw first one ball and then another (you do not put the first back again before you draw the second). Consider the two events

A: the first ball is black
B: the second ball is black

It is then easy to see that $P(A) = 1/2$ and that the conditional probability P(B given A) equals $1/3$ because, if you have removed a black ball in the first draw, there is one out of three left. So far so good. But what about the reversed conditional probability, P(A given B)? In other words, if the second ball is black, what is the probability that the first was also black?

This question seems a little backward. Shouldn't the probability simply be $1/2$? After all, the first ball is already drawn, so how can its color be influenced by the color of the the second ball?

It can't of course, but that is irrelevant. You should not think of the passing of time here and how at the time the first ball is drawn, the color of the second is not even decided yet. Think of the complete experiment to draw one ball first and then a second ball. Let us think about a large number of repetitions of this experiment. First disregard all cases when the second ball is white. For the second ball to be black, we need to get either white/black or black/black. The probability of white/black is $1/2 \times 2/3 = 1/3$, and the probability of black/black is $1/2 \times 1/3 = 1/6$. Add these numbers and find that the probability that the second ball is black is $1/2$. Thus, in a large number of repetitions, about half the time will the second ball be black. In how many of these cases is also the first ball black? As we saw that white/black is twice as likely as black/black, we conclude that black/black accounts for one third of the cases when the second ball is black. We have argued that P(A given B) = 1/3.

Easier yet is not to think too much and just use Bayes' rule, which gives

$$\text{P(A given B)} = \frac{\text{P(B given A)} \times \text{P(A)}}{\text{P(B)}} = \frac{1/3 \times 1/2}{1/2} = 1/3$$

where we computed P(B) = 1/2 above. Note that this calculation was actually an application of the law of total probability (page 28):

$$\begin{aligned}\text{P(B)} &= \text{P(B given A)} \times \text{P(A)} + \text{P(B given (not A))} \times \text{P(not A)} \\ &= 1/3 \times 1/2 + 2/3 \times 1/2 = 1/2\end{aligned}$$

This approach gives another form of Bayes' rule and the one that is typically used in computations. Consider any two events A and B. Suppose that we want to compute the "backward" conditional probability P(A given B) and know how to compute the "forward" conditional probabilities P(B given A) and P(B given (not A)). The calculations we did above give the following version of Bayes' rule:

$$\text{P(A given B)} = \frac{\text{P(B given A)} \times \text{P(A)}}{\text{P(B given A)} \times \text{P(A)} + \text{P(B given (not A))} \times \text{P(not A)}}$$

In the example with the balls, it is not much harder to compute P(A given B) than P(B given A), but there is a big difference in how to think about the two conditional probabilities. The first one, P(B given A), is natural. You draw a ball, note that it is black, and realize that the conditional probability that you get another black ball in the next draw is 1/3. On the other hand, the conditional probability P(A given B) is mysterious. You draw a ball and keep it in our hand. The probability that it is black is 1/2. You draw another ball and notice that it is black. Now the probability that the ball you have in your hand is black is suddenly 1/3 and not 1/2! How weird is that? The ball is there, in your hand, and is not going to change colors just because you drew another black ball!

But the probabilities above are subjective, reflecting how strong your belief is in various events. How strong is your belief that the ball in your hand is black? If you use probabilities to quantify this belief, initially you would say 1/2. Then you draw again and get a black ball. The fact that the second ball is black *changes* your degree of belief in holding a black ball. After all, it is easier to get a black ball in the second draw if you drew a white ball in the first.[1] By computing the conditional probability you now quantify your belief in a black ball as $1/3$. This probability is subjective, but it is not arbitrary; it is based on the objective assessment of what happens in the long run when the experiment is repeated.

One way to think of Bayes' rule is that you can use it to *update* probabilities in the light of new information. In the example above, consider the experiment to draw a ball from the urn with two black and two white balls. The probabilities of drawing a black and a white ball are both $1/2$. In the light of the new information that a second ball has been drawn and is black, you update the probabilities to $1/3$ and $2/3$. In the probabilist's language, the *probability distribution* on the set {black, white} before drawing the second ball was $(1/2, 1/2)$; after drawing the second ball and observing that it was black, the distribution changed to $(1/3, 2/3)$. The two distributions are called the *prior* and the *posterior* distribution.

BAYES AND MY GREEN CARD

Bayes' rule is very useful to evaluate the results of various medical tests. Let me illustrate this with a hypothetical example. Suppose that you are being screened for a disease that occurs with a frequency of about 1% of the population. There are no particular symptoms, and the test is 95% accurate. By this I mean that if you have the disease, the test result is positive with 95% probability and if you do not have it, the test result is negative with 95% probability. If you get a positive test result, how likely is it that you have the disease?

This seems straightforward. If the test is 95% accurate, then there is a 95% chance that you have the disease, right? Nevertheless, let us use Bayes' rule to compute the relevant probability. Let D be the event that you have the disease, and let + denote the event that the test is positive. We know that unconditionally $P(D) = 0.01$ because this is the frequency of the disease. The figure that is given for the accuracy of the test translates into the conditional probability $P(+ \text{ given } D) = 0.95$. You are, however, interested in the conditional probability $P(D \text{ given } +)$, the probability that you have the disease if the test result

[1]It may help to think of a more extreme example. In this case, imagine instead that there are only two balls, one black and one white. If you hold a ball in your hand and it turns out that the other ball is black, you know for certain that your ball is white.

is positive. Note how this is truly backward in a chronological sense because first you either have or do not have the disease, then you test positive. Bayes' rule gives

$$P(\text{D given } +) = \frac{P(+ \text{ given D}) \times P(\text{D})}{P(+ \text{ given D}) \times P(\text{D}) + P(+ \text{ given (not D)}) \times P(\text{not D})}$$

Only P(+ given (not D)) requires a moment's thought. This is the probability to get a positive test result if you do not have the disease so it equals 0.05. Plug all values into Bayes' rule:

$$P(\text{D given } +) = \frac{0.95 \times 0.01}{0.95 \times 0.01 + 0.05 \times 0.99} \approx 0.16$$

so there is only a 16% chance that you have the disease! Even though the test is 95% accurate, you are not very likely to have the disease and may in fact be cautiously optimistic. Bogus calculations? Not at all. You have two conflicting pieces of information to evaluate the risk that you have the disease: (1) It is a fairly unusual disease so you are *not* likely to have it and (2) the test is fairly accurate so you *are* likely to have it. Thus, you have the two numbers 1% and 95% that quantify your risk, and the truth should be somewhere in between. Bayes' rule gives the correct way to combine the two to give a final risk of 16%.

Confusion of the forward conditional probability P(+ given D), which equals 0.95, and the backward probability P(D given +), which equals 0.16, is a common mistake even by those who should be the last to confuse them: A 1978 article in the *New England Journal of Medicine* reports how a problem similar to the one above was presented to 60 doctors at four Harvard Medical School teaching hospitals. Only 11 gave the correct answer, and almost half gave the answer 95%!

Note that the calculations involve the unconditional probability P(D), the frequency of the disease in the population as a whole, the so-called *base rate* (the term *rate* is in this context often used as a synonym to probability). It may seem that this is irrelevant. After all, once you have tested positive and know that the accuracy is 95%, is that not all the information you need? Why would it matter how common or rare the disease is? It does matter for the following reason: The less common the disease, the more likely it is that your positive result is a mistake. Imagine the extreme situation that you are testing with 95% accuracy for a disease that *nobody* has (ovarian cancer among men or prostate cancer among women to mention some silly examples). Then you know for sure that a positive result is in error. The more common the disease, the more you should believe in a positive test result. The other extreme is a condition that *everybody* has; a positive test result is then always true. Base rates cannot be neglected.

In the example, it is very helpful to think about what happens on average if a lot of people are screened. In a population of 10,000, there will be about 100 with the disease and 95 of those will be detected. Among the 9,900 that do not have the disease, 5% will get an erroneous positive test result, which adds another 495 people to the group with positive test results. Thus, we have found 590 positive cases but only 95, about 16%, of these are actually sick. The tests of those who have gotten a positive test result but are healthy are called *false positives*, and the five individuals whose illness was missed by the test represent the *false negatives*. I mentioned that "rate" is often used as a synonym to probability; we thus talk about false-positive rates and false-negative rates. In the example, the false-positive rate is $495/590$ or 84%, and the false-negative rate is $5/9{,}410$ or 0.05%.

To weed out some false positives, a second test can be done on all those who tested positive. Suppose that a second test is done, that its result is independent of the result in the first test, and that this second test also has accuracy 95%. Thus, if you have the disease, the probability that you test positive is 0.95 regardless of how you tested in the first round, and if you are healthy, the probability to test positive is 0.05. The only difference from above is that the base rate is now 0.16 because we are testing in the group of those that tested positive in the first round. If you test positive in the second round, the probability that you are actually sick is

$$\text{P(D given +)} = \frac{0.95 \times 0.16}{0.95 \times 0.16 + 0.05 \times 0.84} \approx 0.78$$

In other words, only 22% false positives remain, a remarkable decrease. In the spirit of the previous section, you may think of the procedure above as sequential updating of probabilities. Your initial risk of 1% was first updated to 16%, then to 78%. These numbers would not be easy to figure out without the help of Bayes' rule.

In any test you obviously want both false-positive rates and false-negative rates to be low, but there is a problem with this because these two goals work against each other. Whatever type of medical test you are doing (temperature, blood tests, X rays, MRI,...), except for the clearcut cases, there are also borderline cases, the gray area where the indication is not clear. If you want to reduce the risk of getting a false positive, you set your levels high to avoid classifying people in the gray area as positive. However, this will lead to more false negatives, those who should be positive but are now cleared. Conversely, if you set your levels low so that you reduce the number of false negatives, you will instead increase the number of false positives. As you have noted, I like to use extreme cases to make my point. Here, consider the extreme test that declares *everybody* positive. As nobody is cleared, this test has a false-negative rate of 0 but there will be a false-positive for everybody who is healthy and thus a potentially high false-positive rate. The other extreme is to declare everybody healthy, which gives no false positives but potentially many false negatives.

There are plenty of real-life situations where false positives and negatives show up. Your engine light comes on, but there is nothing wrong with your car: false positive. Andy Roddick's serve is clearly out, but the linesman misses it: false negative. My spellchecker lets "form" pass when I mistype "I come form Sweden" but reacts to "Olofsson," which is the correct spelling of my last name: one false negative and one false positive.

In medical testing, it is first and foremost desirable to have a low rate of false negatives. The philosophy is that you catch as many affected people as possible. The price to be paid, the increased number of false positives, is then considered less alarming than missing people who are actually affected. However, the increased false-positive rate can also be troublesome. For example, when computerized tomography (CT) scans are used to detect lung cancer, the false-positive rate can be high (sometimes up to 50% according to the National Cancer Institute). A false positive is usually a nodule that turns out not to be cancerous, and this is found out by surgical intervention, a procedure that is at best a great inconvenience and at worst fatal. Although it seems obvious that early detection of lung cancer is a good thing, some researchers advocate caution and point out that many tumors are nonlethal or slow-growing cancers that the patient would likely have died with and not from. The term *screening* is used for testing large groups of people without symptoms but in a risk group such as smokers for lung cancer. There is an ongoing debate over whether screening for early detection of lung cancer is beneficial. There are similar concerns for many types of cancer screenings, such as mammograms, pap smears, and the PSA test for prostate cancer. In general, screenings will typically yield a fairly high false-positive rate, so this has to be weighed against the benefits of early detection.

The accuracy of a medical test for a disease may not be the same for those affected with the disease and for those unaffected. For a simple example, suppose that you use high fever as the only criterion to diagnose measles. If somebody has measles, they will almost certainly have high fever, and the method is in that case almost 100% accurate. However, if somebody does not have measles, there are still plenty of other reasons they may have a fever so the method would not work as well. Many who do not have measles would nevertheless be diagnosed with measles, so the method is much less reliable for those who are afflicted with the disease.

The probability that a test is positive when it should be is called the test's *sensitivity*. This measures how sensitive the test is to detect what it is supposed to. Thus, the fever test for measles has a high sensitivity, which means that it gives very few false negatives (very few cases of measles go undetected). The probability that a test is negative when it should be is called the test's *specificity*. This measures how well the test is able to identify the specifics of the particular disease or condition. As fever is not specific for measles, the fever test has a low specificity and gives many false positives (those who have a fever but do not have measles).

I have some personal experience of false positives. When I applied for my green card, I had to do a skin test for tuberculosis (TB). It came back positive, which is often the case for those of us who come from countries where mass vaccination with the BCG vaccine is done for TB (in Sweden this was done when I was a kid but has since been replaced by selective vaccination). Thus, the specificity of the skin test is reduced by these vaccinations. After a positive skin test, the next step was a chest X ray, which cleared me and I got my green card. It is typical that a first test for a disease has a high sensitivity but lower specificity; the positives will then be tested with other more specific methods. This is the case in HIV testing where the first test is often something called EIA (enzyme immune assay) or ELISA (enzyme-linked immunosorbent assay), which reacts to HIV antibodies but can also react to other antibodies; in which case, it would yield a false positive. A second test is then done, for example, something called a "Western blot," which can tell HIV antibodies apart from other antibodies. Thus, the Western blot has higher specificity than the EIA and ELISA but is more expensive and therefore not used as the initial test.

OBJECTION YOUR HONOR

Bayes' rule can be used to evaluate evidence against a defendant in a court trial. This is very similar to the medical tests we talked about in the last section, with "test positive" replaced by "having evidence against" and "sick" replaced by "guilty." The evidence can, for example, be obtained eyewitness testimony, polygraph tests (if allowed), fingerprints, footprints, or DNA samples. Or, as in the case of Sally Clark from page 22, simply a statement claiming that the deaths of two children were extremely unlikely to have happened naturally.

The idea to use probabilities in court cases is not new. It was discussed already by the great French mathematician Pierre-Simon Laplace who was mentioned early in the first chapter. In his 1814 *Philosophical Essay on Probability*, Laplace describes how Bayes' rule (which at the time had fallen into oblivion and was independently discovered by Laplace) can be used to evaluate evidence in court cases.[2]

Regardless of what the type of evidence is, the fundamental question is whether the defendant is guilty "beyond reasonable doubt." In our terminology, given the evidence, is the conditional probability that the defendant is guilty high enough to motivate a conviction? There are no rules for what is a "high enough" probability, but we can probably agree that it needs to be higher in more serious crimes like when Robert Blake was accused of murder or Mike Tyson of rape, than when Winona Ryder was accused of shoplifting. The reason for this is obvious; the more severe the punishment, the more certain we want to be to convict the right person. A somewhat unexpected consequence of this is that it is easier to get away with rape and murder than with shoplifting.

There is a fundamental difference here from the medical tests. As I mentioned above, in medical tests, it is first and foremost important to have a low rate of false negatives. In court it is more important to avoid a false positive—the conviction of somebody innocent—than a false negative—the acquittal of somebody guilty. The problems that show up are very much the same as in medical testing though: neglect of base rates and confusion of conditional

[2]Laplace's mathematical genius did not always help him. Only six weeks after he had been hired by Napoleon as the Minister of the Interior, he was fired amid complaints that he was looking "everywhere for subtleties" and bringing the "infinitely small" to the affairs of government. An interesting twist is that Napoleon at the age of 16 had Laplace as his teacher and passed his exams.

probabilities forward and backward. Let me start by an example constructed by Kahneman and Tversky (see page 11) regarding eyewitness reports.

Eighty-five percent of the taxi cabs in a city are blue, and the remaining 15% are green. A cab that was involved in a hit-and-run accident at night was identified by an eyewitness as green. Tests showed that the witness was able to tell the difference between green and blue (correctly classify green as green and blue as blue) under similar lighting conditions 80% of the time. What is the probability that the cab was green?

When Kahneman and Tversky presented this problem to people in a research study, the typical answer was 80%. This seems plausible; if the witness is correct 80% of the time, the probability that the witness is correct in this particular case is also 80%. But this fails to take into account the proportion of green and blue cabs in the city. Similar to the introductory example about medical tests on page 98, there are two pieces of evidence regarding the color of the cab: the base rates and the eyewitness report. The base rates point to the cab involved in the accident being blue because blue cabs are more numerous, and the eyewitness report points to the cab being green. As usual, the Reverend Bayes comes to our rescue.

Let the event of interest be G: The observed cab was green. Furthermore, let B be the event that it was blue and E (for "evidence") be the event that the witness has identified the cab as green. We are thus interested in the conditional probability P(G given E). If the cab involved in the accident was in fact green, the probability is 80% that the witness classified it as green. This gives P(E given G) = 0.80, P(E given B) = 0.20, and the base rates are P(G) = 0.15 and P(B) = 0.85. By now you know how to apply Bayes' rule:

$$\begin{aligned} P(\text{G given E}) &= \frac{P(\text{E given G}) \times P(\text{G})}{P(\text{E given G}) \times P(\text{G}) + P(\text{E given B}) \times P(\text{B})} \\ &= \frac{0.80 \times 0.15}{0.80 \times 0.15 + 0.20 \times 0.85} \approx 0.41 \end{aligned}$$

so the probability that the cab involved in the accident was green is 41%. In fact, despite the witness report, the cab is still more likely to have been blue although this probability has dropped to 59% from the initial 85%. Neglect of base rates leads to confusion of the relevant conditional probability P(G given E) and the irrelevant P(E given G). It is easy to understand why people make such mistakes if they have no knowledge of probability or prior exposure to similar situations. There are also those who have argued against Kahneman and Tverksy and claim that base rates are indeed irrelevant in a case like this one. To side with Kahneman and Tversky, consider the following argument. Suppose that we find out that the green cab company had all its cabs at some function in a neighboring city at the time of the accident. Thus, at the time of the accident, *all* cabs in the city were blue. Although the witness has testified that the cab was green and is correct 80% of the time, you can be 100% sure that it is the blue cab company that has to pay the damages. Once again base rates cannot be neglected.

This type of problem is usually presented in such a way that base rates and witness reports point in different directions. It would be interesting to see what people would say if the base rates were reversed. In a more extreme setting, suppose that 99% of all cabs are green and the 80% accurate witness also says green. I suspect most would feel more than 80% certain that the green cab company was liable. You might want to do this with Bayes' rule as an exercise, and you should then get a 99.75% probability that the cab was green. Note that this is not between 80% and 99%; it is larger than both. The initial probability of a guilty green cab is 99%, based on base rates alone. The witness report increases support for a guilty green cab, and the updated probability is higher than 99%.

The cab example illustrates the general problem. Given the evidence, whatever it may be, what is the probability of guilt? Recall the used car dealer from page 28. Suppose that your car indeed develops the kind of engine problems you had feared. Should you return to the dealer to deliver a scolding for unethical business practices? You can always do that out of frustration, of course, but hardly based on your evidence. The probability that your car is flood-damaged given that it has engine problems is $0.80 \times 0.05/0.135 \approx 0.30$ so the dealer is more likely to be innocent than guilty.

For a real case, let us revisit Sally Clark. Recall that she was convicted of murdering two of her children based on an expert witness who claimed that there was only a 1-in-73-million chance to have two cot deaths in the same family, and that seems to have been the only "evidence" against her. The first problem with this figure was that it was based on an unrealistic independence assumption, and we revised the probability to $1/850{,}000$. Let us just round it to one in a million. What does it mean anyway? It means that in a family like the Clarks, the probability that two children will die cot deaths is one in a million, certainly not large. But *any* double infant death is very rare, so regardless of the cause, something very unusual happened in the Clark family. Given the fact that it happened, what is the probability that it was by natural causes and Sally was innocent?

Let us use Bayes' rule again. Let C be the event of double cot deaths and M be the event of a double murder. For simplicity, let us exclude other cases such as one murder and one cot death. The observed event of double deaths can then be written as "C or M" because there are no other possibilities of double deaths. We have been given the probability $\text{P(C)} = 1/1{,}000{,}000$, but the relevant probability—the probability that Sally is innocent—is the conditional probability P(C given (C or M)). As the reversed probability P((C or M) given C) obviously equals one (if you have observed C, you must have "C or M" for certain), Bayes' rule gives the probability of Sally's innocence as

$$\text{P(C given (C or M))} = \frac{\text{P((C or M) given C)} \times \text{P(C)}}{\text{P(C or M)}} = \frac{\text{P(C)}}{\text{P(C)} + \text{P(M)}}$$

where we also used the addition rule from Chapter 1 in the denominator. But now we don't get any further because we don't know what P(M) is. This is the probability that the mother in a family like the Clarks kills two of her children. And whatever this is, it is certainly also very small, undoubtedly much smaller than the probability of two cot deaths. The probability P(M) is very close to zero simply because most mothers do not kill their children, especially not twice. For the sake of argument, let us assume that a double murder is as likely as a double cot death so that P(M) is also equal to one in a million. Insert this above to obtain

$$\text{P(C given (C or M))} = \frac{1/1{,}000{,}000}{1/1{,}000{,}000 + 1/1{,}000{,}000} = 1/2$$

so there is a 50–50 chance that Sally is innocent, which makes sense because something very unlikely has happened and there are two equally as likely explanations. It would be wrong to say that Sally murdered her children just because double cot deaths are so rare because this fails to take into account that double murders are also rare.

If the 1 in 73 million counts as evidence against Sally, what conclusions should we draw about Evelyn Adams from page 72 and her 1 in 17 *trillion*? That it never happened? But it did, and in Sally Clark's case, it is not just unlikely to have two *cot* deaths, it is unlikely to have *any* two deaths. But once this very unlikely event has occurred, we need to evaluate all possible explanations with conditional probabilities given the unlikely event. You can't just choose one alternative and rule it out because it was improbable to start with and rule in favor of another alternative whose initial probability you have not even considered. If Sherlock Holmes had been present at the trial, he would most certainly have opposed the expert witness and his 1 in 73 million:

> You will not apply my precept, he said, shaking his head. How often have I said to you that when you have eliminated the impossible, whatever remains, however improbable, must be the truth.
>
> Sir Arthur Conan Doyle, *The Sign of Four*, 1890

Data on double infanticides in families similar to the Clarks are of course extremely sparse and, I am certain, would yield a probability far lower than that of double cot deaths. If it is, for example, one in ten million, the conditional probability of Sally's guilt becomes $1/11$ or about 9%. I have, by the way, used "cot deaths" and "natural causes" synonymously above. This is not quite correct since a cot death is just one of many natural causes, another fact that speaks in favor of Sally.

In Sally Clark's case, the "evidence" was not specific to her. In most cases, however, the evidence is specific to the defendant. There may be eyewitness reports, fingerprints, footprints, or DNA samples that match the defendant. For a modern legal classic, let us consider the California case *People vs. Collins*. On June 18, 1964, Juanita Brooks was robbed on a street in Los Angeles. An eyewitness saw a blonde woman with a ponytail run from the scene and leave in a yellow car driven by a black man with a beard and a mustache. The police later arrested Janet and Malcolm Collins based on this description. In trial, neither Ms. Brooks nor the witness could identify either defendant and there was no other evidence against the Collinses. To boost its case, the prosecution called in a local math teacher who testified that the probability that a randomly selected couple would have all the reported characteristics (blonde, beard, yellow car, etc.) was 1 in 12 million. He arrived at this figure by assigning probabilities to each of six individual characteristics:

$P(\text{man with mustache}) = 1/4$

$P(\text{black man with beard}) = 1/10$

$P(\text{woman with ponytail}) = 1/10$

$P(\text{blonde woman}) = 1/3$

$P(\text{interracial couple}) = 1/1{,}000$

$P(\text{yellow car}) = 1/10$

and then multiplying them to obtain the probability $1/12{,}000{,}000$. The prosecutor used this figure to claim that there was only a 1-in-12-million chance that the Collinses were innocent. They were convicted.

The defense appealed, and in 1968, the California Supreme Court overturned the guilty verdict on four grounds. Three of these were regarding problems with the calculation of the 1 in 12 million: unfounded probabilities of individual characteristics, incorrect assumptions of independence (for example, between beard and mustache), and neglect of the possibility of incorrect eyewitness reports (including the possibility of false beard, dyed hair etc.). The fourth point is critical though: the fallacy to equate the probability of observing the characteristics with the probability of innocence. Even if the first three problems did not exist and it is true that there is only a 1-in-12-million chance that a random couple has these characteristics, there are millions of couples that could have been at the scene of the crime at that time. And as you know, even as small a p as $1/12{,}000{,}000$ is not sufficient to rule something out if you also have a large n. In this case, it is simply not unlikely enough that there is another couple with the same characteristics to convict based on this piece of evidence alone. Indeed, if we knew for certain that there is another couple that fits the description, there would be a 50–50 chance of the Collinses guilt and it would be impossible to convict. And if we knew that there were two more such couples, the guilt probability would drop to $1/3$, and so on.

Let us leave the Collinses for a while and consider their situation in a more general context. Suppose that some person (or couple, group, corporation, etc.) stands accused of a crime based on only one source of evidence. It is known about this piece of evidence that a match with a randomly selected person occurs with probability p. If there is a match with the defendant, what is the probability of guilt?

This problem is of course yet another case for Reverend Bayes. Let G and I denote guilt and innocence, and let E denote evidence. If the defendant is guilty, then there is by necessity a match with the evidence; if the defendant is innocent, there can still be a match and this occurs with probability p. We therefore have the conditional probabilities P(E given G) $= 1$ and P(E given I) $= p$. The probability that the defendant is guilty given the evidence is therefore

$$\text{P(G given E)} = \frac{\text{P(E given G)} \times \text{P(G)}}{\text{P(E given G)} \times \text{P(G)} + \text{P(E given I)} \times \text{P(I)}}$$

$$= \frac{P(G)}{P(G) + p \times P(I)}$$

and without any idea about the initial probability of guilt, we get no further. Suppose that there are, except for the defendant, n possible suspects that are considered equally as likely to be guilty. This gives a total of $n + 1$ equally likely suspects, and the probability of guilt for any one person is $P(G) = 1/(n + 1)$ which gives $P(I) = 1 - P(G) = n/(n + 1)$, which can be inserted above to give

$$P(G \text{ given } E) = \frac{1}{1 + n \times p}$$

as the probability of guilt given the evidence. This problem is often described in terms of $n + 1$ people on an island, one of whom is a murderer and one of whom happened to match DNA from the scene of the crime. This so-called "island problem" has been debated by lawyers and probabilists (guess who are most likely to get it right).

For the Collinses, p equals $1/12{,}000{,}000$ and n would be of the order of millions. If n is two million, the formula above gives an 86% probability of guilt, which is probably not high enough to convict in the absence of any other evidence. At any rate, there is certainly a big difference between a 14% chance and a 1-in-12-million chance of innocence. In a legal context, the confusion of P(E given I) with P(I given E) is called the *prosecutor's fallacy*. The absurdity of the fallacy can be illustrated by considering less dramatic probabilities than the one in many millions in the Collins case. Suppose for example that a crime has been committed by somebody of blood type AB. This blood type is known to occur with a frequency of about 4% in the U.S. population. If a defendant has blood type AB, the prosecutor's fallacy would be to claim that based on this evidence alone, there is a 96% probability that the defendant is guilty. This is of course ludicrous because 4% translates into almost 12 million people in the United States having the same blood type, and they would thus all be 96% likely to be guilty.

Bayes' rule is also useful in evaluating polygraph tests. First of all, it is often pointed out that the polygraph is *not a lie detector*. OK, we heard that. Now let us call it a lie detector anyway because that is how we think of it. The accuracy of lie detectors is debated. There are websites in support and websites in opposition. I also found one site that gladly exclaims, "Don't worry, the lie detector can be beaten easily!"

Whatever is the truth about lie detectors, the same problems as in evaluation of other types of evidence show up: neglect of base rates and false positives. Suppose that the lie detector is accurate enough to decide "truth" or "lie" with 95% probability. You are testing your normally very honest friend Innocentius and to your surprise get a positive reading. You would initially estimate that there is only a 1-in-a-1,000 chance that he would ever lie. What is the probability that he was actually lying? Let L denote the event that he lied, T that he told the truth, and let + denote a positive reading. The relevant probability is then P(L given +) and Bayes' rule gives

$$\begin{aligned}\text{P(L given +)} &= \frac{\text{P(+ given L)} \times \text{P(L)}}{\text{P(+ given L)} \times \text{P(L)} + \text{P(+ given T)} \times \text{P(T)}} \\ &= \frac{0.95 \times 0.001}{0.95 \times 0.001 + 0.05 \times 0.999} \approx 0.02\end{aligned}$$

so it is still very unlikely that this honorable man has told a lie. As in so many previous cases we have looked at, when you test for something rare with a testing procedure that is not perfect, a positive result is more likely to be a mistake than a true positive.

The most intriguing probabilities come from DNA, evidence. In 1994, a former football player by the name of Orenthal James Simpson stood trial for the murder of...oh, you have heard about it?[3] In the O.J. trial, an expert witness testified that the probability that a blood stain found at the scene of the crime came from somebody other than O.J. was 1 in 170 million. Where does such a figure come from? It does of course not come from going through a DNA data bank including samples from 170 million people; instead, it is actually very similar to what the math teacher did in *People vs. Collins*: multiplication of probabilities. Data are collected from smaller groups of people but from several different *loci* (Latin for "places" and plural of *locus*) of the chromosomes that are known to be independent. For one example, consider blood types. The four blood types according to the "ABO classification" are A, B, AB, and O. Which type you are is determined by a particular gene on one of your chromosomes. In addition, blood is also classified by the Rh factor, which can be positive or negative. This is determined by a gene on another chromosome and is therefore independent of your ABO type. For example, 40% of the U.S. population are known to have blood type A and 16% are

[3]Incidentally, this crime took place only a few miles from the scene of the crime in the Collins case and to the day 30 years later on June 18, 1994. What are the odds...?

Rh-negative. By independence we can conclude that 16% of 40%, that is, 6.4%, are A-negative.

Similarly with "DNA fingerprinting." Several loci are investigated in terms of frequencies of different genetic patterns, and by multiplying these frequencies, the probability of a random match can be calculated from relatively small samples (but not too small because then the natural variation between individuals would be missed). In the O.J. Simpson case, the figure 1 in 170 million was computed from a sample of 240 African-Americans in Detroit. It is important to realize that different subpopulations are different genetically. Thus, in the O.J. case, it would not have been very helpful if the probability calculations had been based on DNA samples from Amish in Millersburg, Ohio or Cajuns in Lafayette, Louisiana. The 1 in 170 million seems compelling but was discredited by accusations that DNA evidence had been tainted by the police and by some admitted mistakes in the probability calculations. At any rate, the probabilistic nature of DNA evidence makes it difficult to grasp for the untrained, and one juror said after the trial "I didn't understand the DNA stuff at all." He then went on to say "I wish there would have been a probabilist in court to explain it so that I could better understand and incorporate the DNA evidence in my judgment." If only! What he actually said was, "To me, it was just a waste of time. It was way out there and carried no weight with me."

The appearance of probabilists in court is not unprecedented. In the 1996 British rape trial *Regina vs. Adams*, Oxford professor Peter Donnelly gave the jury a tutorial on Bayes' rule and how to apply it to evaluate evidence. In this case, the only evidence against the defendant was a DNA match, which was reported to have a 1-in-200-million probability. There were several pieces of evidence in the defendant's favor, including an alibi given by his girlfriend and the failure of the rape victim to identify him. Mr. Donnelly explained to the jurors how they could use probability calculations and Bayes' rule to come up with a final probability of guilt (a number that could be different for different jurors). I will not go into the details of the case but only mention that Adams was convicted, the case appealed, and the conviction upheld by the Appeal Court. A good account of the case is given by Peter Donnelly himself in a 2005 article in the journal *Significance*.

When dealing with the extremely low probabilities involved in DNA evidence, it is important to realize that they apply to the population as a whole (or the particular subpopulation that is under consideration). Thus, the probability that a randomly chosen individual matches the DNA may be 1 in 200 million, but if you test a *relative* of the defendant, the probability of a match is much higher. Indeed, if the defendant has an identical twin, his or her DNA is identical,[4] which is something that would not be incorporated in the "1 in 200 million." And siblings, parents, cousins, aunts, and uncles are also genetically much closer than randomly chosen people, and thus, they match the defendant's DNA with higher probabilities. In *Regina vs. Adams*, this was an issue because the accused had a half-brother whose DNA was for some reason never tested.

I think we can agree that probability arguments in court are both important and difficult. After the appeal of the Adams case, the Appeal Court issued the following statement regarding the use of Bayes' rule:

> We regard the reliance on evidence of this kind in such cases as a recipe for confusion and misjudgment, possibly even among counsel, but very probably among judges and, as we conclude, almost certainly among jurors.

I would object and claim that a proper use of Bayes' rule is actually a recipe to *avoid* confusion and misjudgment. That is, of course, merely the opinion of one humble probabilist who, if ever put on trial, would insist that a jury "of his peers" were to be composed entirely of other probabilists.

FINAL WORD

Among mathematical results, Bayes' rule may be the one that has traveled the farthest from an inconspicuous beginning to a wide and still growing range of modern applications. There is a whole school of statistical methodology that is often referred to as *Bayesian statistics* and at the bottom of this rests, of course, Bayes' rule, although usually in more complicated versions than what we have encountered in this chapter. This particular approach to statistics involves very heavy and complex calculations, and it was not until the recent advent of fast and sophisticated computers that it started to gain ground. In this chapter, we have focused on the use of Bayes' rule in "probabilistic detective work," how to draw conclusions based on evidence in court cases, medical trials, and other similar situations. The key here is that the interest is in backward calculation:

[4]Although identical twins have identical DNA, the DNA can be *expressed* in different ways which is why it is still possible to tell identical twins apart (after some practice). One amusing consequence is that it is harder to tell identical twins apart by DNA fingerprinting than by regular old fingerprints obtained by ink and paper.

We can see the outcome but wonder what led to it. Given a particular scenario, we can figure out the probability to get the outcome, but our goal, to figure out the probability of the scenario given the outcome, can only be reached by the use of Bayes' formula.

5

Beyond Probabilities: What to Expect

GREAT EXPECTATIONS

In the previous chapters, I have several times talked about what happens "on average" or what you can "expect" in situations where there is randomness involved. For example, on page 86, it was pointed out that the parameter λ in the Poisson distribution is the *average* number of occurrences. I have mentioned that each roulette number shows up on *average* once every 38 times and that you can *expect* two sixes if you roll a die 12 times. The time has come to make this discussion exact, to look beyond probabilities and introduce what probabilists call the *expected value*. This single number summarizes an experiment, and in order to compute an expected value, you need to know all possible outcomes and their respective probabilities. You then multiply each value by its probability and add everything up. Let us do a simple example.

Roll a fair die. The possible outcomes are the numbers 1 through 6, each occurring with probability $1/6$, and by what I just described, we get

$$1 \times 1/6 + 2 \times 1/6 + 3 \times 1/6 + 4 \times 1/6 + 5 \times 1/6 + 6 \times 1/6 = 3.5$$

as the expected value of a die roll. You may notice that the term "expected" is a bit misleading because you certainly do not expect to get 3.5 when you roll the die. Think instead of the expected value as the expected *average* in a large number of rolls of the die. For example, if you roll the die five times and get the numbers 2, 3, 1, 5, 3, the average is $(2+3+1+5+3)/5 = 2.8$. If you roll another five times and get 2, 5, 1, 4, 5, the average over the ten rolls is $31/10 = 3.1$. As you keep going, rolling over and over and computing consecutive averages, you can expect these to settle in toward 3.5. I will elaborate more on this interpretation and make it precise in the next chapter. You can also think of the "perfect experiment" in which the die is rolled six times and each side shows up exactly once. The average of the six outcomes in the perfect experiment is 3.5, and this is the expected value of a die roll.

In the casino game *craps*, two dice are rolled and their sum recorded. What is the expected value of this sum? The 11 possible values are 2, 3,..., 12, but these are not all equally likely so we must figure out their probabilities. In order to get 2, both dice must show 1 and the probability of this is $1/36$. In order to get 3, one die must show 1 and the other 2 and as there are two dice that can each play the role of "one" or "the other," there are two outcomes that give the sum 3. The probability to get 3 is therefore $2/36$. To get 4, any of the three combinations 1–3, 2–2, or 3–1 will do, so the probability is $3/36$, and so on and so forth. The outcome 7 has the highest probability, $6/36$, and from there the probabilities start to decline down to $1/36$ for the outcome 12 (consult Figure 1.2 on page 6 if you feel uncertain about these calculations). Now add the outcomes multiplied by their probabilities to get

$$2 \times 1/36 + 3 \times 3/36 + \cdots + 12 \times 1/36 = 7$$

as the expected sum of two dice. This time, the expected value is a number than you *can* get as opposed to the 3.5 with one die. Unfortunately, it still does not mean that you actually expect to get 7 in each roll or you would expect to leave the casino a wealthy person, 7 being a winning number in craps. It only refers to what you can expect on average in the long run.

Note that the expected value of the sum of two dice, 7, equals twice the expected value of the outcome of one die, 3.5. This is no coincidence. Expected values have the nice property of being what is called *additive*, which means that we did not have to do the calculation we did above for the two dice. Instead, we could just have said that as we roll two dice and each has the expected value 3.5, the expected value of the sum is $3.5 + 3.5 = 7$. This is convenient. If you roll 100 dice, you know that the expected sum is 350 without having to figure out how to combine the outcomes of 100 dice to get the sum 298 or 583 (but feel free to try it, at least it will keep you out of trouble).

Expected values are more than additive; they are also *linear*, which is a more general concept. In addition to additivity, linearity means that if you multiply each outcome by some constant, the expected value is multiplied by the same constant. For example, roll a die and double the outcome. The expected value of the doubled outcome is then twice the expected value of the outcome of the roll, $2 \times 3.5 = 7$. Note that the expected values are the same when you *double* the outcome of *one* roll and when you *add* the outcomes of *two* rolls. The actual experiments are different though. In the first case, the possible values are the 6 even numbers 2, 4,..., 12; in the second, the 11 numbers 2, 3, ..., 12.

To illustrate the convenience of linearity, suppose that you construct a random rectangle by rolling three dice. The first determines one side, and the sum of the other two determines the other side. What is the expected circumference of the random rectangle? There are 216 different outcomes of the three dice. The smallest possible rectangle measures one by two, has circumference 6, and probability $1/216$ because there is only one way to get it: (1,1,1). The largest rectangle measures 6 by 12, has circumference 36, and likewise probability $1/216$. In between these, there is a range of possibilities with different probabilities. For example, you can get circumference 8 in three different ways: (1,1,2), (1,2,1), and (2,1,1), so circumference 8 has probability $3/216$ (these three rectangles have dimensions 1×3, 1×3, and 2×2, respectively). To compute the expected circumference, however, you do not need to figure out all these outcomes and their probabilities. Simply note that the circumference is twice one side plus twice the other; that the sides have expected lengths 3.5 and 7, respectively; and apply linearity to get the expected circumference $2 \times 3.5 + 2 \times 7 = 21$. Linearity is a very convenient property indeed.

You may have noticed that the expected values in our examples thus far have been right in the middle of the range of possible outcomes. The midpoint of the numbers 1, 2,..., 6 is 3.5; the midpoint of 2, 3,..., 12 is 7; the midpoint of 100, 101,..., 600 is 350; and the midpoint of the rectangle circumference values is $(6 + 36)/2 = 21$. All these examples have in common that the probability distributions are *symmetric*. If you roll one die, start from 3.5, which is in the middle and step outward in both directions: 3 and 4 have the same probabilities; 2 and 5 have the same probabilities; and 1 and 6 have the same probabilities. Of course, in this particular case, *all* outcomes have the same probabilities, $1/6$, so it may be more interesting to look at the sum of two dice instead. Here, 7 is in the middle and has probability $6/36$. One step out we find 6 and 8, which both have probability $5/36$. Continue like this until you hit the last two outcomes 2 and 12, each with probability $1/36$. So in these cases, you could actually have found the expected value simply by computing the average of the possible outcomes: The average of 1, 2,..., 6 is 3.5, and the average of 2, 3,..., 12 is 7. This is not always the case though, and here is another dice example to prove it. Roll two dice and record the *largest* number. What is the expected value?

The largest number can be anything from 1 to 6, but these are not equally likely, nor are the probabilities distributed symmetrically, and it is probably clear that the largest value is expected to be more than 3.5. To find the expected value, we need to first compute the probabilities. The only case in which the largest number equals 1 is when both dice show 1, and this has probability $1/36$. Three cases give the largest number 2: 1–2, 2–1, and 2–2, and the probability is thus $3/36$. Continue like this until you reach 6, which is the largest number in 11 cases and has probability $11/36$ (you might again find it helpful to consider Figure 1.2). If you are into math formulas, the probability that the largest number equals k is $(2 \times k - 1)/36$ for k ranging from 1 to 6. At any rate, the expected value is now computed as

$$1 \times 1/36 + 2 \times 3/36 + \cdots + 6 \times 11/36 \approx 4.5$$

rounded to one decimal. I leave it up to you to demonstrate that the expected *smallest* number is ≈ 2.5 (obvious without calculations?).

Here is another example of asymmetric probabilities, which also involves negative numbers in a natural way. You play roulette and bet $1 on the number 29. What is your expected gain? There are two possibilities: With probability $1/38$, number 29 comes up and you win $35, and with probability $37/38$, some other number comes up and you lose your dollar. If we agree to describe a loss as a negative gain, your gain can therefore be either 35 or -1. There is no problem with having a negative number, and the expected value of your gain is computed just like before:

$$35 \times 1/38 + (-1) \times 37/38 = -2/38 \approx -0.0526$$

an expected loss of about 5 cents. Again we have a case in which the expected value cannot actually occur but must be interpreted as a long-term average. In the long run, each number comes up once every 38 spins, so assume that this is *exactly* what happens; the numbers come up perfectly in order: 00, 0, 1, 2, ..., 38 and you bet $1 on 29 each time, wagering a total of $38. You will then lose \$37 and win \$35 (and keep the dollar you bet on 29), a total loss of \$2 out of the \$38.

You often see people at the roulette tables betting on several different numbers, sometimes covering almost the entire table. Although this certainly increases your chances of winning in a single spin, it does nothing to improve your expected long-term losses. Indeed, you lose 5 cents per dollar on each single number, so if you for example bet \$1 on each of ten different numbers, additivity of expected values tells you that you can expect to lose on average 50 cents. Regardless of betting strategy, the casino takes on average 5 cents out of every dollar you risk, which does not sound like much but is enough to give them a very good profit.

As an exercise, let us compute the expected gain in the more innocent chuck-a-luck. In this game, you wager \$1, three dice are rolled, and your win depends on the number of 6s. If there is one 6, you win \$1; if there are two 6s, you win \$2; and if there are three 6s, you win \$3. Only if there are no 6s do you lose your \$1. On page 18, we saw that the probability that you win something is 0.42. Thus, you lose your \$1 with probability 0.58, but if you win, you may win more than \$1 so it is not immediately obvious that the game is stacked against you. The probabilities to get zero, one, two, and three 6s are

$$\begin{aligned}
P(\text{no 6s}) &= (5/6)^3 &&= 125/216 &&\approx 0.58\\
P(\text{one 6}) &= 3 \times 1/6 \times (5/6)^2 &&= 75/216 &&\approx 0.35\\
P(\text{two 6s}) &= 3 \times (1/6)^2 \times 5/6 &&= 15/216 &&\approx 0.07\\
P(\text{three 6s}) &= (1/6)^3 &&= 1/216 &&\approx 0.005
\end{aligned}$$

where the "3" in the two middle probabilities is there because the one die that shows different from the other can be any of the three (in fact, the number of 6s has a binomial distribution, which we discussed in Chapter 1). The decimal numbers above do not add up to 1 as they should, but that is only because they are rounded. Let us now compute the expected gain. Your gain equals the number of 6s if you get any, and otherwise, it is -1. The expected gain in chuck-a-luck is

$$(-1) \times 125/216 + 1 \times 75/216 + 2 \times 15/216 + 3 \times 1/216 \approx -0.08$$

that is, an expected loss of about 8 cents per \$1 wagered. From a financial point of view, you're worse off than at the roulette table. Again you can think of what would happen in the ideal run where each of the 216 possible outcomes of the three dice comes up exactly once. You then win \$3 once, \$2

15 times, \$1 75 times, and lose \$1 125 times, for a total loss of \$17 of your \$216 wagered.

Suppose that you try another kind of gambling: stock investments. A friend tells you that a particular mutual fund is equally likely to either go up 50% or down 40% each year for the next few years to come. If you invest \$1,000, how much can you expect to have after two years?

First consider one year. After the first year you are equally likely to have \$1,500 or \$600, and the average of these is \$1,050. In general, the average of a 50% gain and a 40% loss is a 5% gain, so you can expect to gain 5% each year. After two years, your expected fortune is therefore $\$1,000 \times 1.05 \times 1.05 =$ \$1,102.50. On the other hand, as your fortune is equally likely to increase as it is to decrease each year and there are two years, you can expect it to go down one year and up the other. Regardless of which of these years that comes first, your fortune will be $\$1,000 \times 1.50 \times 0.60 = \$900$. This seems conflicting. How can you expect your fortune both to increase and to decrease?

It depends on what you mean by "expect." The expected value of your fortune after two years is certainly \$1,102.50. There are four equally likely scenarios for the two years: up–up, up–down, down–up, and down–down, leading to fortunes of \$2,250, \$900, \$900, and \$360, respectively, and the average of these is \$1,102.50. However, if you instead compute the expected number of "good years," this number is one and the *most likely* scenario is one good and one bad year, which makes \$900 the most likely value of your fortune. The most likely value, in this case \$900, is called the *mode* or *modal value*. It is up to you which of the two measures of your fortune you think makes most sense. Note that although your *expected* fortune increases, the *actual* fortune only increases if there are two good years of which there is a 25% chance. If you compare this investment scheme with one that gives a fixed 5% interest each year, the two are on average equally good and equally likely to be ahead after a year. However, the fixed interest scheme has a 75% chance of being ahead of the mutual fund after two years. If they compete, it is a fair game year by year but not over two years, somewhat paradoxically. And as the years keep passing by, your expected fortune increases by 5% each year, but under the most likely scenario your fortune instead decreases by 10% every two years. After 20 years, your initial investment of \$1,000 has grown to \$2,653 as measured by expected returns and fallen to \$349 under the most likely scenario. In order for your actual fortune to increase after 20 years, you need at least 12 good years, which has a probability of about 25%.

The rates in the example may not be very realistic but serve as a drastic illustration to the general principle that a decrease is more severe than an increase. For example, if a 50% gain is followed by a 50% loss (or vice versa), this leaves you with a net loss of 25%. The combination of a 10% gain and a 10% loss results in a net loss of 1%, and so on. If equally sized annual gains and losses are equally likely, your expected fortune remains unchanged, but in order for the actual fortune not to decrease, you need more good years than bad. This is still true even if the annual gains tend to be slightly larger than the annual losses (as in the extreme example above).

When it comes to risking money in order to make money, you must of course weigh risk against benefit and considering only the expected gain is not sufficient. You may buy a lottery ticket for the slim chance to win big even though you face an expected loss. But if I offer you the chance to bet $1,000 on a coin toss and pay you $1,100 if you get heads, you might not want to play even with the expected gain of $50. In the long run you would certainly ruin me, but for a single bet you might not be willing to risk $1,000 for the chance of winning that extra $100. You face similar concerns when it comes to investing your money. Should you take a risk on highly uncertain but potentially very profitable stocks, or should you go with the lower risks of mutual funds or bonds? The expected return should play a role in your decision but should definitely not be the sole criterion.

Careful consideration of expected values can save time and money as the next example illustrates. During World War II, millions of American draftees had their blood drawn to be tested for syphilis, a disease that was expected to be detected in a few thousand individuals. Analyzing the blood samples was a time-consuming and expensive procedure, and a Harvard economist, Robert Dorfman, came up with a clever idea. Instead of testing each individual, he suggested, divide the draftees into groups, draw their blood, and mix some blood from everybody in the group to form a *pooled* blood sample. If the pooled sample tests negative, the whole group is declared healthy, and if it tests positive, each individual sample is tested separately. The point is of course that entire groups can be declared healthy by just one blood sample analysis. The same idea can be used for any disease that is rare and where large populations need to be screened. Let us look at the mathematics of pooled blood samples.

Denote the size of the group by n and the probability that an individual has the disease by p.[1] Additional tests must be done if *somebody* has the disease, and because *somebody* is the opposite of *nobody*, this is a case for Trick Number One. The probability that an individual does not have the disease is $1 - p$, and assuming independence between individuals, the probability that nobody has the disease is $(1 - p)^n$. Finally, the probability that somebody has the disease is $1 - (1 - p)^n$, and this is then the probability that the pooled sample tests positive; in which case, n additional individual tests are done. After the first test, with probability $(1 - p)^n$, there are no additional tests, and with probability $1 - (1 - p)^n$, there are n additional tests. The expected number of tests with the pooling method is therefore

$$1 + n \times (1 - (1 - p)^n)$$

where the first 1 is there because one test must always be done and the term $0 \times (1 - p)^n$ that should formally be added was ignored because it equals 0. Now compare this expected value with the n tests that are done if all samples are tested individually. Let us put in some values, for example, $n = 20$ and $p = 0.01$. Then $1 - p = 0.99$, and the expected number of tests is

$$1 + 20 \times (1 - 0.99^{20}) \approx 4.6$$

which is certainly preferred over the 20 tests that would have to be done individually. Note also that even if the pooled blood sample is positive, very little is lost because the pooling method then requires a total of 21 tests instead of 20, only one test more (and there is no need to draw more blood, what was drawn initially is used for both pooled and individual tests). The probability of a positive pooled blood sample is $1 - 0.99^{20} \approx 0.18$, so if people are divided into groups of 20, about 18% of the groups need to undergo the individual testing. One practical concern is that if groups are too large, the pooled blood sample might become too diluted and single individuals who are sick may go undetected. In the case of syphilis, however, Dorfman points out that the diagnostic test is extremely sensitive and will detect the antigen even in very small concentrations. Dorfman's original article, bearing the somewhat politically

[1] Epidemiologists use the term *prevalence* for the proportion of individuals with a certain disease or condition. For example, a prevalence of 25 in 1,000 for us translates into $p = 0.025$. A related term is *incidence*; this is the proportion of *new* cases in some specific time-period.

incorrect title "The detection of defective members of large populations" was published in 1943 in the *Annals of Mathematical Statistics*. The procedure of pooling has many applications other than blood tests, for example, tests of water, air, or soil quality.

Let me finish with a little treat for the theory buffs. First of all, the expected value is commonly denoted by μ (Greek letter "mu"). The general formula for μ is as follows. If the possible values are $x_1, x_2, \ldots$, and these occur with probabilities $p_1, p_2, \ldots$, respectively, the expected value is defined as

$$\mu = x_1 \times p_1 + x_2 \times p_2 + \cdots$$

where the summation goes on for as long as it is needed. In the case of a die roll, the summation stops after six terms, x_k equals k and all p_k equal $1/6$. For another example, recall the binomial distribution from page 39. This counts the number of successes in n independent trials where each time the success probability is p. The possible outcomes are the numbers 0, 1, ..., n, and the corresponding probabilities were given in the formula on page 39. The expected number of successes is therefore

$$\mu = \sum_{k=0}^{n} k \times \binom{n}{k} \times p^k \times (1-p)^{n-k}$$

which is not completely trivial to compute. However, it is easy to guess what it is. For example, if you toss a coin 100 times, what is the expected number of heads? Fifty. If you roll a die 600 times, what is the expected number of 6s? One hundred. In both cases, the expected number is the product of the number of trials and the success probability, and this is true in general. Thus, the binomial distribution with parameters n and p has expected value $n \times p$ (which as usual for expected values is not necessarily a possible actual outcome). If you are familiar with Newton's binomial theorem, you might be able to show that the expression for μ above indeed equals $n \times p$.

GOOD THINGS COME TO THOSE WHO WAIT

There are expected values where the summation in the formula from the previous section goes on forever. This does not mean that it takes forever to compute them, only that we can get an infinite sum if there is no obvious limit on the number of outcomes. For example, if you toss a coin repeatedly and count how many tosses it takes you to get heads for the first time, this number can theoretically be any positive integer. Although it is highly unlikely that you have to wait until the 643rd toss, you cannot rule it out. There is thus an infinite number of outcomes. I have already pointed out that probabilists do not fear the infinite, and our notation for the expected value in a case like this is

$$\mu = \sum_{k=1}^{\infty} x_k \times p_k$$

where ∞ is the infinity symbol, indicating that the sum never ends. It is one of the little intricacies of higher mathematics that you can add an infinite number of terms and still end up with a finite number. The probabilities p_k must of course eventually become very, very small. The probability that you get your first head in the 643rd toss is, for example, $(1/2)^{643}$, which starts with 193 zeros after the decimal point. In general, the probability that you get the first head in the kth toss is $(1/2)^k$, and the expected number of tosses until you get heads is

$$\sum_{k=1}^{\infty} k \times (1/2)^k = 1 \times (1/2) + 2 \times (1/2)^2 + 3 \times (1/2)^3 + \cdots$$

and, believe it or not, this messy expression equals 2. This is intuitively appealing though. As heads show up on average half the time, they appear on average every other toss and your expected wait ought to be two tosses. By changing the success probability from $1/2$ to $1/6$, an even messier sum can be shown to equal six; thus, the expected wait until you roll a 6 with a die is six rolls. And yet another change, to $1/38$, reveals that each roulette number is expected to show up once every 38 spins. In general, if you are waiting for something that occurs with probability p, your expected wait is $1/p$. One of the rewards of studying probability is that mathematics and intuition often agree in this way. Another reward is of course that math and intuition do often *not* agree, at least not immediately, thus yielding wonderfully surprising results. As you have already learned, probability certainly has a complex and contradictory charm.

There is another way to compute the expected wait until the first heads, or 6, or roulette win, a way that avoids the infinite sum. Recall how we, starting on page 29, computed winning probabilities in some racket sport problems by considering a few different cases, one of which led back to the starting point, thus giving an equation for the unknown probability. We can use such a *recursive* method here too. Suppose that you wait for something that occurs with probability p and let μ denote the expected wait. In the first trial, you either get your event of interest or you do not. If you do, the wait was one trial. If you do not, you have spent one trial and start over with an additional expected wait of μ trials, yielding a total of $1 + \mu$ expected trials. As the first case has probability p and the second $1 - p$, you get the following equation for μ:

$$\begin{aligned}\mu &= p \times 1 + (1 - p) \times (1 + \mu)\\ &= 1 + \mu - p \times \mu\end{aligned}$$

which simplifies further to the equation $0 = 1 - p \times \mu$ that has solution $\mu = 1/p$, just like we wanted.

Let us look at a variant of the problem of waiting for something to happen. In the *Seinfeld* episode "The Doll," Jerry is very happy to find a dinosaur in a cereal box (right after Elaine has told him he is juvenile). Let us now say that there are ten different plastic toys to be found in that type of cereal box. In order to get all of them, what is the expected number of boxes Jerry must buy? This is difficult to solve directly by using the definition of expected value. In order to do this, you would have to compute the probability that it requires k boxes for the possible values of k, and as there is no upper limit on these values, this presents a tricky problem. Just try to compute the probability that Jerry must buy 376 or 12,971 boxes.

We will do something smarter. First of all, one box is bought and contains a dinosaur. What is the expected number of boxes Jerry must buy in order to get a different toy? As the probability to get a different toy is $9/10$, he can expect to buy $10/9$ boxes, in analogy with what I said above with $p = 9/10$. Once he has gotten two different toys, he starts waiting for one different from these, and as there are now eight remaining toys, the probability to get something different is $8/10$ and the expected number of boxes is $10/8$. Next, he can expect to buy another $10/7$ boxes, then $10/6$, and so on until he finally can expect to buy $10/2 = 5$ boxes to get the second-to-last toy and $10/1 = 10$ boxes to get the final toy. Finally, in order to get the expected number of boxes Jerry must buy to get all the toys, we use the additivity property of expected values and conclude that he can expect to buy

$$1 + 10/9 + 10/8 + \cdots 10/2 + 10/1 \approx 29$$

boxes. Note that one third of these are bought in order to get the very last toy, every parent's nightmare. The expression above can be rewritten in a mathematically more attractive way as

$$10 \times \left(1 + \frac{1}{2} + \frac{1}{3} + \cdots + \frac{1}{9} + \frac{1}{10}\right)$$

where the expression in parenthesis consists of the first 10 terms of the *harmonic series*. It is a well-known mathematical result that as more and more terms are added, the harmonic series summed up to n terms, H_n, gets close to the natural logarithm of n, denoted $\log n$ (or sometimes $\ln n$). The natural logarithm of a number x is what the number e (= 2.71828..., remember the discussion on page 85) must be raised to in order to get x. Thus, if $e^y = x$, then y is the natural logarithm of x: $y = \log x$.[2] As n increases, the difference $H_n - \log n$ approaches a number that is known as *Euler's constant* and is approximately equal to 0.58.[3] We can now establish a nice general formula for the expected number of cereal boxes if there are n different toys:

$$\text{expected number of boxes} \approx n \times (\log n + 0.58)$$

which for $n = 10$ gives 28.8, approximately 29 just like above. If n is very large, we need to refine the constant 0.58; see footnote 3. This type of problem did not start with Jerry Seinfeld; it is a classic probability problem usually called the *coupon collecting problem* and has been generalized in a multitude of ways.

A related type of problem is the so-called *occupancy problem*. If Jerry learns that he can expect to buy 29 boxes in order to get all the toys and decides to go on a cereal shopping spree and buy 29 boxes at once, how many

[2]A more familiar logarithm is the base-10 logarithm where e is replaced by 10. For example, the base-10 logarithm of 100 is 2 because this is what 10 must be raised to in order to get 100: $10^2 = 100$. In a similar way, you can consider the logarithm in any base and, for example, conclude that the base-4 logarithm of 64 is 3 because $4^3 = 64$. The ancient Babylonians liked the base 60 and we still use this to keep track of time in seconds, minutes, and hours. In our everyday math, we use base 10, computer scientists like the bases 2 (binary system) and 16 (hexadecimal system), but to mathematicians only the base e is worthy of consideration.

[3]Leonhard Euler, a Swiss mathematician who lived between 1707 and 1783, was one of the greatest mathematicians ever. He was extremely prolific and contributed to almost every branch of mathematics. His collected works fill over 70 volumes, and his name has been given to so many mathematical results that when you refer to "Euler's Theorem" you have to specify *which* Euler's theorem that you are talking about. The constant mentioned here has an infinite decimal expansion starting with 0.5772156..., following no discernible pattern, and it is a famous unsolved problem whether it is *rational* (can be written as a fraction of two integers) or *irrational*.

different toys can he expect to get? Note that it is *not* ten. Of course he could be really unlucky and get dinosaurs in all of them, which has probability $(1/10)^{29}$. Multiply this probability by 10 and get the probability that only one type of toy (not necessarily a dinosaur) is represented. It is also possible to have 2, 3, ..., 9, or 10 different toys represented in the 29 boxes. The expected number of toys must be somewhere between 1 and 10, but it is tricky to compute directly. Again, additivity comes to our rescue and in this case in a really clever way.

Open all of the boxes, and first look for dinosaurs. If you find any, count "one" and otherwise "zero" (note that you count "one" if you find *at least* one dinosaur; you do *not* count the *number* of dinosaurs). Next, look for another type of toy, say, a SAAB 900 (a car model featured in several *Seinfeld* episodes). If you find any, count "one" and otherwise "zero." Keep looking for other types of toys, each time counting "one" if you find it and "zero" otherwise. When you have done this ten times, you have ten ones and zeros, and if you add them, you get the number of different toys that are represented (if your sum equals ten, they are all there). Thus, to get the final number, you added ten numbers and by additivity of expected values, to get the expected final number, you add ten expected values, each such expected value being computed from something that can be either one or zero. Also note that all these individual expected values are equal because there is no difference between the toys in regard to how likely they are to be in the box. What then is such an expected value?

In order to find it, we only need to figure out the probabilities of "one" and "zero." If the probability of 1 is p, the probability of 0 is $1 - p$ and the expected value is

$$0 \times (1 - p) + 1 \times p = p$$

and by adding ten such expected values, we realize that the expected number of different toys is $10 \times p$. To find p, note that we count "one" if there is at least one dinosaur. The ever useful Trick Number One tells us that the probability of this is one minus the probability of no dinosaurs, and we get

$$\text{P(at least one dinosaur)} = 1 - (9/10)^{29}$$

and finally

$$\text{expected number of different toys} = 10 \times (1 - (9/10)^{29}) \approx 9.5$$

so the juvenile Jerry is quite likely to get all his toys.

Let us summarize the coupon collecting problem and the occupancy problem in a general setting. There are n different types of objects and you are attempting to acquire them one by one. The expected number of attempts until you get all of the n objects is

$$n \times \sum_{k=1}^{n} (1/k) \approx n \times (\log n + 0.58)$$

and if you try N times, the expected number of different objects that you get is

$$n \times \left(1 - \left(\frac{n-1}{n}\right)^N\right)$$

where you can notice that this number is very close to n if N is large, as you would expect.

The zeros and ones that you summed above are called *indicators* because they indicate whether a certain type of toy is present in the boxes. Resorting to indicators is a very useful technique to compute expected values, another example being the *matches* that we discussed on page 88. For a quick reminder, if you write down the integers 1, 2, ..., n in random order, the probability that there are no matches (no numbers left in their original position) is approximately 0.37 regardless of n. It is also possible to compute the probability of one match, two matches, and so on, and from this we could compute the expected number of matches. However, to find the expected number of matches, it is easier to use indicators. Simply go through the sequence and count one whenever there is a match and zero otherwise. Add the zeros and ones to get the number of matches. To get the expected number of matches, we only need to figure out the probability of a match in a particular position and multiply this by n, just like with the toys in cereal boxes above. This is easy. Focus on a particular position. As the numbers are rearranged at random, the probability that this position regains its original number is simply $1/n$ and the expected number of matches is therefore $n \times 1/n = 1$. Regardless of how many men leave their hats at the party, when hats are randomly returned, one man is expected to get his own hat back.

EXPECT THE UNEXPECTED

In the previous chapters we have seen many examples where probability calculations lead to results that are surprising or counterintuitive. This is the case for expected values as well, and we will look at several examples. First, some random geometry.

Suppose that you create a random square by rolling a die to determine its sidelength. You then also compute the area, which is the square of the sidelength. The possible sidelengths are thus 1, 2,..., 6; the possible areas are 1, 4,..., 36; and each sidelength S corresponds to precisely one area A according to the equation $A = S^2$. Plain and simple. Let us now compute the expected sidelength and area. The expected sidelength is easy; we already know that this is 3.5. For the expected area, we can then square this value and get $3.5^2 = 12.25$. Or can we? Better be careful and do the formal calculation. As each sidelength has probability $1/6$ and corresponds to exactly one area, each area also has probability $1/6$ and we get the expected area

$$1 \times 1/6 + 4 \times 1/6 + \cdots + 36 \times 1/6 \approx 15.2$$

which is not at all 12.25. Apparently we cannot just square the expected sidelength to get the expected area. This becomes clearer if we think about long-term averages. For example, occurrences of sidelengths 1 are in the long run compensated for by sidelengths 6 and they average 3.5. However, when you compute the corresponding areas, sidelength 1 gives area 1 and sidelength 6 gives area 36; these areas average 18.5, which is not the square of 3.5. In the same way, sidelengths 2 and 5 average 3.5, but the corresponding areas 4 and 25 average 14.5. When all areas are averaged, in the long run, the average will settle around 15.2. Notice that this number is *higher* than the square of the expected sidelength. This is because areas grow faster than sidelengths; doubling the sidelength quadruples the area. So when you say that "the average square has sidelength 3.5 and area 15.2," it may sound absurd but of course you will never actually see the "average square."

Here is a simple game. You and a friend are asked to take out your wallets and count your cash. The only rule of the game is that whomever has more must give it to the other (and if you have exactly the same amount, nothing happens). Would you agree to play this game? You might argue: "I know how much money I have. If my opponent has less, I lose what I have and if he has more, I win more than what I have. There is no specific reason to believe that he is poorer or wealthier than I am, so this seems like a good deal. In fact, since I have just learned about expected values, let me try to compute my expected gain. My x dollars can lead to either a loss of x dollars or a gain of y dollars, where $y > x$ and since a gain and a loss each have probability $1/2$, my expected gain is

$$(-x) \times 1/2 + y \times 1/2 = (y - x)/2$$

which is always a positive amount."

The math formula looks impressive, and you no longer hesitate but conclude that the game is in your favor, and you accept to play. However, when you see the smug look on your opponent's face, you suddenly realize that he has gone through similar calculations and come to the conclusion that the game is in *his* favor, so he is also eager to play. This makes you confused. How can the game be favorable to *both* of you?

The paradox stems from your implicit assumption that you are equally likely to win or lose, regardless of the amount in your wallet (that is where the probability $1/2$ comes from). Clearly this is not true. For example, if you have no money at all, you are almost certain to win unless your opponent is also broke. At least you cannot lose anything. If you have some, but very little money, you are quite likely to win, but if you have a lot of cash, chances are that your opponent has less and you lose. Remember, "either/or" is not the same as "50–50."

Let us look at a simple example. Suppose that you and your opponent simply flip a coin each to decide how much cash you have. Heads means you have \$1, tails that you have \$2. If you and your opponent flip the same, nothing happens. If you flip heads and he flips tails, you win \$1; if you flip tails and he flips heads, you lose \$1. As these two scenarios are equally likely, your expected gain is \$0 and the game is fair.

OK, that was easy. Let us make it a little more complicated and suppose instead that you and your opponent choose your cash amounts by each rolling a die. What is your expected gain? First, we can ignore all ties. Second, there is a certain inherent symmetry in that, for example, the outcome (3,5) (your amount first) has the same probability as the outcome (5,3). In the first case you win \$2; in the second, you lose \$2. In this fashion, each gain is canceled by an equally probable loss of the same size, and as you sum over all possible outcomes, you end up with \$0 and the game is again fair.

Now, people don't go around and toss coins or roll dice to decide how much cash they have. But these were only examples to illustrate that we can describe the amount of money in a wallet at some arbitrary time as generated by some random mechanism. There is an amount of uncertainty in numbers and sizes of cash withdrawals and cash payments, and in the end, it is reasonable to assume that there is a range of possible cash amounts to which we can ascribe probabilities. It is fairly easy to show (and even easier to believe) that the expected gain for each player is $0, regardless of what this range and these probabilities are, as long as they are the same for both players.

One of the first to describe the wallet paradox was Belgian mathematician Maurice Kraitchik in his 1942 book *Mathematical Recreations*, but with neckties instead of cash. I found it in Martin Gardner's 1982 book *Aha! Gotcha*, a collection of various mathematical puzzles. Mr. Gardner does not seem to have fully grasped the problem though. In his own words, "We[4] have been unable to find a way to make this clear in any simple manner" and points out that Kraitchik himself "is no help." But Mr. Gardner also remarks that the paradox perhaps arises because each player "wrongly assumes his chances of winning or losing are equal," and as I explained above, this is precisely the resolution to the paradox. As I mentioned in Chapter 2, Mr. Gardner pursued a lifelong devotion to educating the general public in mathematics, and considering this noble task, let us forgive him his somewhat indecisive treatment of the wallet paradox.

The wallet paradox was puzzling at first, but I think we managed to eventually set it straight. The next paradox is similarly mindboggling and not so easy to resolve. You are presented two envelopes and are told that one contains twice as much money as the other. You choose an envelope at random, open it, and note that it contains $100. You are now asked if you want to keep the money or switch and take what is in the other envelope. First, there does not seem to be anything to gain from switching, but then you start thinking. The other envelope contains $50 or $200, and since you chose randomly, it is equally likely to be either. Thus, by switching you either gain $100 or lose $50, and your expected gain is

[4]You may have noticed that mathematicians are very fond of the *pluralis majestatis*, a manner of expression traditionally reserved for royalty. Mark Twain proposed to extend the privilege to people with tapeworms; mathematicians seem to have added themselves to the list. Personally I believe this is because mathematicians are a very friendly and communal minded bunch who often feel that manipulating math formulas is a lonely business.

$$(-50) \times 1/2 + 100 \times 1/2 = 25$$

so it seems to be to your advantage to switch.

OK, so switch then, what is the problem? Well, there is nothing special with the amount \$100, and the calculations can be repeated for any amount A that you find in the first envelope; in which case, the other envelope contains $A/2$ or $2 \times A$ and your expected gain is

$$(-A)/2 \times 1/2 + 2 \times A \times 1/2 = A/4$$

dollars. Thus, it is always to your advantage to switch, so why even bother opening the first envelope? Just take it and immediately switch to the other. But why even bother taking the first? Just take the other envelope directly! But wait, then that envelope has become the first so shouldn't you then switch to the other, formerly first, envelope? But then you should take that envelope directly instead. But then...

Now that was really confusing. Something must be wrong but what? Let us try to do the experiment and see what happens. We get two envelopes, put two amounts of money in them, and start choosing, opening, and switching. What will happen? Naturally, you win as often as you lose in the long run, and the amount you win or lose is always the same. There are *two* envelopes and *two* amounts of money, but above we had *three* possible amounts floating around: $A/2$, A, and $2 \times A$. Even though you may observe A dollars in your envelope and have no reason to believe more in either of the amounts $A/2$ and $2 \times A$ in the other, it does not seem sensible to translate this into probabilities the way we did above. Once again, "either/or" is not necessarily the same as "50–50." In this case, it is actually either "0–100" or "100–0," you just do not know which.

A better description is that you are presented two envelopes that contain A and $2 \times A$, respectively, for some amount A. If you choose at random, open and switch, you are equally as likely to gain \$A as you are to to lose \$A. The world makes sense again, and the envelope problem is not fun anymore.

SIZE MATTERS (AND LENGTH, AND AGE)

Consider a randomly sampled family with children. On average equally as many boys are born as girls; therefore, such a family has, on average, equally as many sons as daughters. But this must mean that boys tend to have more sisters than brothers. For example, in a family with four children, the average composition is two sons and two daughters, and in such an average family, each boy has two sisters but only one brother. On the other hand, once a boy is born, the rest of the children should be born in the usual 50–50 proportions, which indicates that boys tend to have equally as many brothers and sisters. What is correct?

The second claim is correct. Boys to *not* tend to have more sisters than brothers. This may seem paradoxical at first, though. If you sample a boy at random and he has on average the same number of brothers as sisters, once you add him to the mix does this not indicate that there tend to be more boys than girls in the family? Yes indeed, but there is a twist. There is a difference between sampling a *family* and sampling a *boy*. Indeed, when you sample a boy, you are ruling out the families that have only girls, always selecting a family that has at least one son, and *such* a family *does* on average have more sons than daughters. For a simple illustration, consider only families with two children so that the equally likely gender combinations listed by birth order are GG, GB, BG, and BB. If a family is sampled at random, the probability that it has no sons is $1/4$, the probability that it has one son is $1/2$, and the probability that it has two sons is $1/4$. The expected number of sons in the family is therefore

$$0 \times 1/4 + 1 \times 1/2 + 2 \times 1/4 = 1$$

but if a *boy* is sampled at random, the number of sons *in his family* (himself included) is equally likely to be one or two, the reason being that you are now choosing from the four Bs, two of which are paired with a G and the other two with another B. The expected number of sons is therefore

$$1 \times 1/2 + 2 \times 1/2 = 1.5$$

When the sampled boy is removed, the remaining expected 0.5 sons just means that his sibling is equally likely to be male or female. Thus, the average family has exactly one son who still manages to have on average half a brother (not a half-brother, mind you). But just like "average square" earlier, "average family" is not a precise concept unless we specify how the sampling is done. You may also think about it like this: Suppose that children from 1,000 families are gathered at a meeting. There will then be roughly the same number of boys and girls present. Suppose instead that 1,000 *boys* are gathered and that each has brought all his siblings. In the entire group, there will then tend to be more boys than girls present, but among the siblings of the selected boys, proportions are still 50–50. Boys do not tend to have more sisters than brothers; rather, they tend to belong to families that have more sons than daughters.

If you did not get this right the first time, you are in good company. Our constant companion Sir Francis Galton noticed in his 1869 book *Hereditary Genius* that British judges were all men and came from families that had on average five children. He erroneously concluded that the judges therefore had on average 2.5 sisters and 1.5 brothers. Thirty-five years later he realized his mistake and corrected it in an article with the intriguing title "Average number of kinsfolk in each degree" published in the journal *Nature* in 1904 (following an even more intriguingly entitled article, "The forest-pig of Central Africa" by zoologist Philip L. Sclater).

When we sample a boy or a British judge rather than a family, this is an example of *size-biased sampling*. Let us take a closer look at the two-children family. If a family is sampled at random and the number of boys counted, this number can be 0, 1, or 2, and the corresponding probabilities are $1/4$, $1/2$, and $1/4$. In the terminology from page 98, the probability distribution on the set $\{0, 1, 2\}$ is $(1/4, 1/2, 1/4)$. Now instead sample a boy. The probability distribution on the same set is then instead $(0, 1/2, 1/2)$, and the interesting thing is that these new probabilities can be obtained by multiplying each of the first three probabilities by its corresponding outcome: $0 = 0 \times 1/4, 1/2 = 1 \times 1/2$, and $1/2 = 2 \times 1/4$. In other words, the probabilities changed proportional to size: 0 boys became 0 times as likely, 1 boy as likely as before, and 2 boys twice as likely. The new probability distribution is therefore called a *size-biased distribution*.

For another example, roll a die. The set of possible outcomes is then the set $\{1, 2, 3, 4, 5, 6\}$ where each outcome has probability $1/6$. Rather than rolling the die, you can think of this as choosing a face of the die at random. Now instead choose a face of the die by first choosing a *spot* at random, and then choosing the face that this spot is on. As there are $1 + 2 + \cdots + 6 = 21$ spots, the probability to get the face showing 1 is $1/21$, the probability to get the face showing 2 is $2/21$,..., and the probability to get the face showing 6 is $6/21$. The probability distribution on the same set $\{1, 2, 3, 4, 5, 6\}$ is now $(1/21, 2/21, ..., 6/21)$ instead of the distribution $(1/6, 1/6, ..., 1/6)$ we get when we choose a face at random. If we follow the idea in the previous example with the two-children family and multiply each outcome with its corresponding probability in the old distribution, we get $(1 \times 1/6, 2 \times 1/6, ..., 6 \times 1/6)$, that is,

$(1/6, 2/6, ..., 6/6)$. This set of numbers is not a proper probability distribution because the sum of the numbers is not equal to one. However, if each number is multiplied by $6/21$, we get precisely the new distribution when a spot is chosen at random. Again, the probabilities in the new distribution have changed by a factor proportional to size. The new size-biased probability of k is the old probability $1/6$ multiplied by $6 \times k/21$.

There is more to be said. As $21/6 = 3.5$, which is the expected value of a die roll, the size-biased probability is in fact the old probability times the size of the outcome divided by the expected value, $1/6 \times k/3.5$. Let us look at this more formally. Denote the old probability of k by p_k, the expected value by μ, and the size-biased probability by $\widehat{p}_k$. We then have the relation

$$\widehat{p}_k = k \times p_k/\mu$$

for $k = 1, 2, ..., 6$. In our particular case, the p_k are all equal to $1/6$ and $\mu = 3.5$, but the relation we stated between the p_k and the $\widehat{p}_k$ is true for any probability distribution on any set. The size-biased distribution is the old distribution with each probability multiplied by k/μ.

For another example of size-biased sampling, suppose that you choose a U.S. state by randomly sampling and recording the state of (a) a U.S. Senator and (b) a member of the U.S. House of Representatives. Then (a) is equivalent to choosing a state at random, whereas (b) is size-biased sampling because larger states have more House representatives and are thus more likely to be chosen. If you want all states to be equally likely, choosing a member of the House is incorrect, but if you want to give more weight to more populous states, it is correct. In general, size-biased sampling may be something you do not wish to do and that happens by mistake, but it may also be precisely what you want to do. There are many real-life situations where some type of size-bias becomes an issue. When an individual is chosen at random for an opinion poll, she is likely to come from a family that is larger than average, live in a city that is larger than average, go to a school that is larger than average, work for a company that is larger than average, and so on, all of these being factors that may have an impact on her opinions. When an ichthyologist catches fish, this may be done by detecting an entire school and larger schools are easier to detect. The same situation arises for any kind of animal that appears in clusters, be it flocks of birds, armies of frogs, or smacks of jellyfish. When a forest is inspected from the air for a tree disease, larger patches of sick trees are easier to detect. Larger tumors are easier to detect on a scan or X ray. And so on and so forth; size definitely matters.

Now let our randomly chosen family take a trip to Yellowstone National Park where the most visited attraction is the *Old Faithful* geyser, famed for its regular eruptions, which occur about every 90 minutes. When our friends arrive, they would thus expect to wait 45 minutes for an eruption. As they wait, they start talking to a man who has visited many times and has carefully recorded his waiting times, which average more than 45 minutes. He tells our family that this indicates that the geyser is slowing down, but data from the park rangers do not give such indications. Other than that our family's new friend may have had some bad luck, is there a logical explanation?

Definitely. The crux is that the Old Faithful, contrary to her name and reputation, does not erupt *exactly* every 90 minutes, only on average. Indeed, times between eruptions vary between 30 minutes and 2 hours but are most typically in the 60–100-minute range or so. If it did erupt exactly every 90 minutes and you arrived at a random time, your expected wait would certainly be 45 minutes. But now that intervals vary in length, you are in fact more likely to arrive in one of the longer intervals and thus your expected wait is longer than 45 minutes. To simplify things, suppose that intervals alternate between one and two hours so that eruptions occur at noon, 2 P.M., 3 P.M., 5 P.M., 6 P.M., and so on. The average interval length is then 90 minutes, but if you arrive at random, you are twice as likely to arrive in a 2-hour interval and your expected wait is one hour; if you arrive in a 1-hour interval, your expected wait is half an hour. Thus, two thirds of the time you wait on average an hour and one third of the time, half an hour. As $2/3 \times 1 + 1/3 \times 1/2 = 5/6$, your expected wait is $5/6$ of an hour or 50 minutes, longer than half the average interval time 45 minutes. See Figure 5.1 for an illustration of this scenario. In reality there is of course much more randomness than just shifting back and forth between one- and 2-hour intervals but you get the general picture.

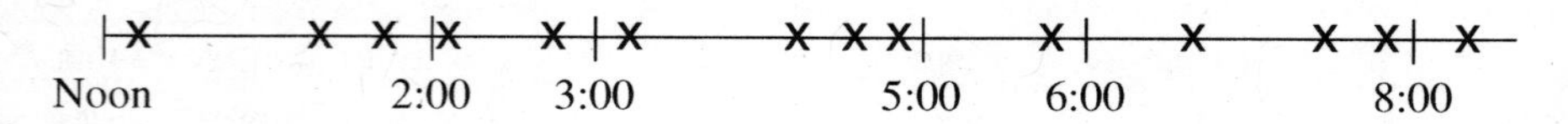

Figure 5.1 The Old Faithful erupting at alternating intervals of lengths one hour and two hours and successive random arrivals. Note that there are more arrivals in the 2-hour intervals, making the average waiting time for an eruption more than 45 minutes.

The situation described above is an example of the *waiting time paradox*, a well-known phenomenon in probability. Another example of the waiting time paradox is when you catch a bus by randomly arriving at a bus stop. Even though the bus may run on average twice an hour, due to random variation, you are more likely to hit the longer intervals and must wait on average more than the 15 minutes' waiting time you would have if they ran exactly every half-hour. However, bus arrivals are still fairly regular and the difference is not likely to be large. It is not until the case of the rare and unpredictable events we studied in Chapter 3 that the name "paradox" is really earned. Let us look at earthquakes as an example. According to the U.S. Geological Survey, great earthquakes (magnitude 8 and higher on the Richter scale) occur on average once a year worldwide. Considering the capricious nature of earthquakes, let us agree that they qualify as rare and unpredictable. But this means that at any given time, the expected waiting time until the next great earthquake is one year, regardless of when the previous earthquake occurred, so if a space alien decides to pay a surprise visit to Earth, he can expect to wait one year for the next earthquake. On the other hand, when he arrives, the expected time since the *last* earthquake is also one year (just think of time running backward). One year since the last earthquake, one year until the next, yet one year between earthquakes and not two! Seems paradoxical but remember that these are expected values, and our alien friend is simply more likely to arrive in an interval that is longer than usual. Very short intervals that contribute to lowering the expected length are likely to be missed completely.

The waiting time paradox has a lot in common with size-biased sampling. Consider, for example, the simplified Old Faithful example with intervals between eruptions that are equally likely to be one hour or two hours. A randomly sampled interval is then equally likely to be of either length, and its expected length is 90 minutes. However, when you *arrive* at random, you can think of this as sampling an interval where the 2-hour interval is twice as likely as the 1-hour interval. Thus, the initial probability distribution $(1/2, 1/2)$ on the set $\{30,60\}$ (minutes) has changed to $(1/3, 2/3)$, where more weight is given to the larger value. Note how the new probabilities are proportional to the old probabilities times the interval lengths. Thus, the new distribution is size-biased or, more appropriately in this case, *length-biased*. This was the simplified example but regardless of what the real distribution of inter-eruption times are, when you arrive at random you choose such an interval with a probability proportional to its length.

A similar type of bias shows up when *life expectancy* is computed. In our terminology, life expectancy is the expected lifespan of a newborn individual. In a human population, life expectancy is estimated by recording the ages of everybody who dies (usually in a year) and taking the average. In the *Seinfeld* episode "The Shower Head," George Costanza tries to convince his parents to move to Florida by pointing out that life expectancy in Florida is 81 and in Queens where they live, 73. Does this mean that Frank and Estelle could expect to live eight years longer in Florida? Not quite. One reason (other than the orange juice) that Florida has a high life expectancy is that many people move there from other states, most notably New York. As these people have already started their lives, and in most cases lived a good part of it, they cannot die at an age lower than that of their move. Thus, they "deprive" Florida of deaths at a young age, and this increases the average age at death. This scenario is typical for any city, state, or nation that has net immigration, another well-known (likewise orange cultivating) example being Israel. At the other end, states with net emigration have lower life expectancies. To help you understand, consider an extreme example and suppose that people born in A-town die either at age 40 or at age 80. They live and work in A-town, and if they survive age 40, they retire at 65 and then move to B-town where they live the rest of their lives. Life expectancy in A-town is 40 and in B-town 80, even though the people are really the same. Introduce a more realistic variability in lifespans and migration ages and you get a less drastic but similar effect.

DEVIANT BEHAVIOR

Let us once again sit down at the roulette table.[5] Other than betting on a single number, there are plenty of ways to bet on a whole group of numbers. On the roulette table, the numbers 1–36 are laid out in a 3 by 12 grid where the top row is 1–2–3, the second row 4–5–6, and so on. Also, half of these numbers are red and half are black. On top of this grid are the numbers 0 and 00, colored green (on American roulette tables; European tables do not have the double zero). To bet on a single number is called a *straight bet*. You can also, for example, place an *odd bet*, which does not mean that you are betting in an unusual manner but that you win if any of the odd numbers 1, 3,..., 35 comes up. Likewise, you can bet on even or on red or black. You can

[5]I am constantly whetting your appetite with little glimpses into the world of gambling. Be patient. In Chapter 7, we will indulge shamelessly in all kinds of games, bets, and gambles.

also do *split bets, street bets, square bets, column bets*, and yet some. This is casino lingo, and all it means is that you can place your chip so that it marks more than one number and you then win if any of your numbers come up. Needless to say, the amount you win is smaller the more numbers you have chosen and the payouts are carefully calculated so that you lose 5 cents per $1 regardless of how you play. For example, let us say that you wager your dollar on an odd bet. The payout of such a bet is $1, and since there are 18 odd numbers between 1 and 36, the probability that you win $1 is 18/38 and with probability 20/38 you lose your wagered dollar. Your expected gain is therefore

$$1 \times 18/38 + (-1) \times 20/38 = -2/38 \approx -0.05$$

an expected loss of 5 cents per $1, just like if you place a straight bet. With the odd bet, your chances of winning are significantly higher than with the straight bet, but when you win, the payout is much smaller. In other words, the variability of your fortune is much greater when you place straight bets. This fact is not reflected in the expected value, so it would be nice to have a way to measure variability, in other words, to measure how much the *actual* value tends to differ from the *expected* value. There are different ways to do this, but probabilists and statisticians have come to the consensus that the best measure of variability is something called the *variance*. This is defined as the expected value of the square of the difference between the actual value and the expected value.[6] That was a mouthful. Let me illustrate it with the roulette example of odd bets. The expected value of your gain is -0.05 (dollars) and the two possible actual values are -1 and 1. The differences from the expected value are $-1 - (-0.05) = -0.95$ and $1 - (-0.05) = 1.05$ respectively. Square these two values to get $(-0.95)^2 = 0.9025$ and $1.05^2 = 1.1025$. Finally, we need to compute the expected value of these squared differences. As the first of them corresponds to a loss, it has probability 20/38 and the second, corresponding to a win, has probability 18/38. This gives the variance as the

[6]Squares are computed because we want to have only positive values. Another way to achieve this would be to compute *absolute values* of the differences between actual and expected values (i.e., the differences without signs). It turns out that squares have nicer mathematical properties than absolute values; for example, with some restrictions, variances are additive just like expected values, something that would not be true if we had used absolute values instead of squares.

expected value of these two squares:

$$0.9025 \times 20/38 + 1.1025 \times 18/38 \approx 1$$

a number that in itself does not mean much, but let us compare with the straight bet. Here, the possible actual values are -1 and 35 and a similar calculation to the one above gives a variance that is approximately 33. The much larger value of the variance of the gain of a straight bet than that of an odd bet reflects the larger variability in your fortune with the straight bet. In the long run, you lose just as much with either type of bet, but the paths to ruin look different.

The variance thus supplements the expected value in a useful way. Let us look at another example, the inexhaustible conversation topic of weather. Two U.S. cities that for different reasons caught my attention in early 2006 were Arcata and Detroit. In January 2006, I visited Arcata on the coast of northern California. Browsing through some weather statistics, I calculated that the daily high temperatures have an annual average of about 59 degrees Fahrenheit. A few weeks later, Super Bowl XL was played in Detroit, which has the same annual average daily high of about 59 degrees. Choose a day of the year at random to visit Arcata or Detroit, and the expected daily high is the same, 59 degrees. However, this does not mean much until it is also supplemented with the variance, which for Arcata is 12 and for Detroit 363 (and I challenge you to find a place with a lower temperature variance than Arcata). The much larger variance of Detroit reflects the larger variability in temperatures over the year. For example, the average daily high in Detroit in January is 33 and in July, 85. The corresponding numbers for Arcata are 55 and 63. In Detroit you will need to bring shorts or long johns depending on the season; in Arcata, none of these garments are of much use (but bring an umbrella in the winter).

I mentioned in passing above that there is no clear meaning of the value of the variance. One problem is that it is computed from values that have been squared, which means that the units of measurement have also been squared. What does it mean that the variance is 33 square dollars or 363 square degrees? Nothing, obviously, but there is an easy fix: Compute the square root of the variance. This number is called the *standard deviation* and is more meaningful because the unit of measurement is preserved. In the roulette example, the standard deviations for straight and odd bets are \$1 and $\sqrt{33} \approx \$5.7$, respectively. In the weather example, the standard deviation for Arcata is 3.5 degrees and for Detroit, 19 degrees.

This feels a little better, but the standard deviation still does not have the crystal clear interpretation that the expected value has. There are some rules and results that can help, one of them due to another great Russian mathematician, Pafnuty Lvovich Chebyshev, who lived between 1821 and 1894 and is famous for his contributions to probability, analysis, mechanics, and, above all, number theory.[7] His result, known as *Chebyshev's inequality*, states that in any experiment, the probability to get an outcome within k standard deviations of the expected value is at least $1 - 1/k^2$, for any value of k. For example, choosing $k = 2$ informs us that regardless of what the experiment is, the probability to get an outcome within two standard deviations of the expected value is at least 0.75. Stated differently, Chebyshev's inequality tells us that in a set of observations, at least 75% of the observations fall within two standard deviations of the average. In Arcata, we can expect at least 273 days with a daily high temperature between 52 and 66, and in Detroit, we can expect at least 273 days between 21 and 97 degrees. And with $k = 3$, we get $1 - 1/k^2 = 8/9 \approx 0.89$; at least 89% of observations are within three standard deviations of the expected value.

I would like to stress the "at least" part of Chebyshev's inequality. In reality the probabilities and percentages are often significantly higher. For example, in the roulette example with odd bets, *all* observations are within two standard deviations. Also note that if you choose $k = 1$, all Chebyshev tells you is that at least 0% of the observations are within one standard deviation. Certainly true but not very helpful. Chebyshev's inequality tends to be crude in this way but that is only natural because it is always true, regardless of the particulars of the experiment. It is sort of like saying that every U.S. state is smaller than 572,000 square miles in area. This is needed to include Alaska and is certainly true if we only consider the continental United States, but then 262,000 square miles would be enough. And if we restrict ourselves to New England, even less is needed. Despite these shortcomings, Chebyshev's inequality is still useful as we will learn in the next section.

Let me again pander to those of you who suffer from theory cravings and give the formal definition of variance. Suppose that our experiment can result in the outcomes $x_1, x_2, \ldots$, and that these occur with probabilities $p_1, p_2, \ldots$,

[7]Chebyshev also holds the unofficial world record among mathematicians for most spellings of last name. I should really say transliterations rather than spellings because in his native Cyrillic alphabet he is Чебышёв and nothing else. In the Western world, he has appeared in print in about a dozen different forms ranging from the minimalist Spanish version *Cebysev* to the consonant-indulgence of the German *Tschebyscheff*.

the same setup as we had when we formally defined expected value earlier. Denote this expected value by μ, and remember that the variance involves computing the squared differences between each possible value and μ, then computing the expected value of these squared differences. Translating this verbal description into mathematics gives the formal definition of the variance, commonly denoted by the symbol σ^2 (square of the Greek letter "sigma") as

$$\sigma^2 = (x_1 - \mu)^2 \times p_1 + (x_2 - \mu)^2 \times p_2 + \cdots$$

where the summation stops eventually if there are a finite number of outcomes and goes on forever otherwise. Check for yourself that this is precisely what we did above in the roulette examples. Just for practice, let us do the variance for the roll of a die. The possible values are 1, 2,..., 6, each with probability $1/6$, and the expected value is 3.5. The variance is therefore

$$(1 - 3.5)^2 \times 1/6 + (2 - 3.5)^2 \times 1/6 + \cdots + (6 - 3.5)^2 \times 1/6 \approx 2.9$$

which gives a standard deviation of 1.7. Let us compare this with the standard deviation of a die that has 1 on three sides and 6 on the remaining three. This die gives 1 or 6, each with probability $1/2$, so it also has an expected value 3.5. Its variance is

$$(1 - 3.5)^2 \times 1/2 + (6 - 3.5)^2 \times 1/2 = 6.25$$

which gives standard deviation 2.5. This is larger than the standard deviation of the ordinary die because this special die has outcomes that tend to be further away from the expected value 3.5. Again, we have an example where the expected value does not tell the full story but is nicely supplemented by the standard deviation.

Recall that the standard deviation is the square root of the variance, and it is therefore denoted by σ and we can state the formal version of Chebyshev's inequality. Before we do that, though, let me mention an important concept in probability. Before any experiment, the outcome is unknown and we can denote it by X, which means that X is unknown before the experiment and gets a numerical value after. Such an unknown quantity whose value is determined by the randomness of some experiment is called a *random variable*. This is a very important concept in probability that greatly simplifies the notation in many examples. If a die is rolled, instead of writing things like "the probability to get 5" and "the probability to get 6," we can first denote the outcome of the die by X and write $\mathrm{P}(X = 5)$ and $\mathrm{P}(X = 6)$, a mathematical and more convenient notation. Chebyshev's inequality can now be stated as

$$\mathrm{P}(\mu - k \times \sigma \leq X \leq \mu + k \times \sigma) \geq 1 - 1/k^2$$

or, using absolute values,

$$\mathrm{P}(|X - \mu| \leq k \times \sigma) \geq 1 - 1/k^2$$

for any value of k (which by the way does not have to be an integer; it could be 1.5 or 4.26 or any other nonnegative number). Make sure that these last two expressions are equivalent, and that they agree with the verbal description of Chebyshev's inequality that I gave earlier.

FINAL WORD

The concept of expected value that we have investigated in this chapter can be thought of as the ideal average in a random experiment. The expected value summarizes the experiment in a single number, but we have seen many examples of how some care must be taken in the interpretation of this. The expected value's constant companion is the standard deviation that measures the amount of variability in the experiment, and together the two, μ and σ, provide a convenient summary of the random experiment. I have also at times hinted that we can interpret the expected value as the long-term average, and in the next chapter, this particular interpretation will be thoroughly investigated.

6

Inevitable Probabilities: Two Fascinating Mathematical Results

ALEA IACTA EST, OVER AND OVER

In the beginning of Chapter 5, I mentioned how consecutive averages of repeated die rolls settle in toward the expected value 3.5 as the number of rolls increases. When I wrote this, I expected you to accept it as reasonable or perhaps, if you are the suspicious type, try it out for yourself. And early in Chapter 1, I gave an interpretation of the probability of heads being 0.5: that you can expect to get heads about 50% of the time in a large number of coin tosses. The *proportion* of heads settles in toward the *probability* of heads. It seems that both probabilities and expected values can be interpreted in this way in terms of average long-term behavior, at least in these examples. It is easy to believe that such an interpretation is always possible, and it is certainly nice to have our theory solidly anchored in the real world in this way. Now that we have become sophisticated, this kind of vague and intuitive reasoning is not enough though. We want proof, unquestionable mathematical proof, that averages approach expected values and that proportions approach probabilities. Luckily, this can be done.

That is, it can be done if we can make some reasonable assumptions, for example, that the coin is not tossed by probabilist and magician Persi Diaconis who has the unusual ability to make a coin land heads every time (and can

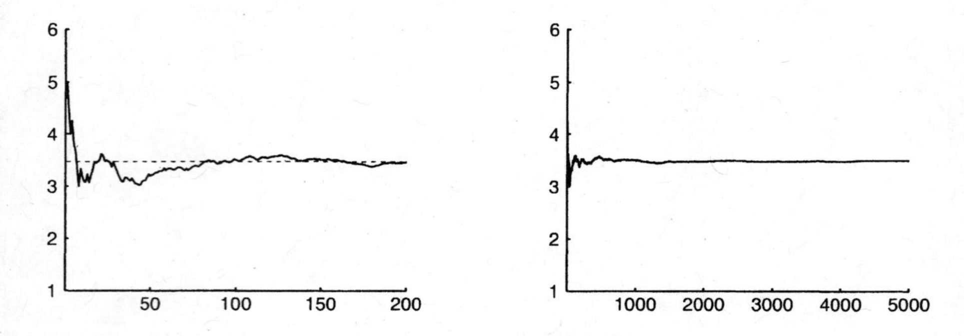

Figure 6.1 Consecutive averages of 200 (left) and 5,000 (right) simulated rolls of a fair die. The expected value 3.5 is marked by a dashed line.

probably make it disappear too).[1] But once we have adequately described the experiment and assigned probabilities to the various outcomes, we are in the world of probability, mathematics, and logic, and we can then prove what is the most fundamental result in probability: the *law of large numbers*, also popularly known as the *law of averages*.

Before proceeding to a formal description, let me illustrate how the law of averages works in an experiment. Figure 6.1 shows consecutive averages computed in a series of rolls of a fair die (computer simulated). The very first roll gave 5, so the average after one roll is simply 5. The second roll gave 3, so the average of the first two rolls is $(5 + 3)/2 = 4$; the third roll gave 3, which makes the third average $(5 + 3 + 3)/3 \approx 4.3$; next followed 4 and 3, which brought the following two averages down to 3.75 and 3.6, respectively, and so on and so forth. It is remarkable how rapidly the averages settle in toward 3.5. There are some fluctuations in the beginning, which is typical. The first roll is equally likely to be any of the numbers 1 through 6 so what happens in the very beginning is highly unpredictable, but the deviations quickly become small. It can, for example, be computed that the average of 100 rolls is between 3 and 4 with probability more than 99.5%.

[1]Diaconis is a legendary figure in probability circles. At the age of 14, he dropped out of school and ran away from home to join a traveling magician, and after years of magic tricks and exposure to various casino games, he became interested in probability. At age 24, he started taking evening classes at the College of the City of New York, still doing magic in the daytime, and five years later, he had earned a Ph.D. degree from Harvard University. He is now a professor at Stanford University and one of the world's most original and prolific researchers in probability and statistics. Take warning though: That's about as far as you can get without a high-school diploma.

As you can see in the figure, there is a small but noticeable decline around roll 180. By pure chance, between rolls 160 and 180, there was an unusually large number of ones and twos, which dragged the average down. In the right plot in Figure 6.1, the sequence is continued to 5,000 rolls. The dip around 180 is now barely visible. If another sequence of 20 rolls similar to that between 160 and 180 would occur after 5,000 rolls, it would have no visible impact.

You have seen an example of how consecutive averages stabilize at the expected value. The counterpart for probabilities is how proportions, or *relative frequencies*, of an event stabilize at the probability of the event. Let us consider the favorite example, coin tosses. Suppose that we toss a coin five times and get the sequence T T H T H. The first five relative frequencies of heads are then 0, 0, $1/3$, $1/4$, and $2/5$ or in decimal form 0, 0, 0.33, 0.25, and 0.4.[2] There is nothing special about having probability 0.5. If we roll a die and consider consecutive relative frequencies of 6s, these approach $1/6 \approx 0.17$ as the number of rolls gets large. Figure 6.2 shows consecutive relative frequencies of heads in 100 coin tosses (left) and consecutive relative frequencies of 6s in 100 die rolls (right). In both cases, note how there are some fluctuations in the beginning but how the relative frequencies stabilize quickly around 0.5 and 0.17, respectively. The path to stabilization looks a bit different for the die roll because it tends to experience longer sequences without success (remember that we must wait on average six rolls between consecutive 6s). This is reflected in the longer "downhill slopes" in the graph. It is, for example, not unusual to experience sequences of 10 or 15 rolls without any 6s, and this happened between rolls 20 and 30 and between rolls 40 and 55. Similar long sequences of coin tosses without heads occur much less frequently.

An imaginative, although not very practical, application of the law of averages was suggested by French eighteenth-century naturalist and scientist Count de Buffon.[3] Consider a floor made of parallel strips of wood, one inch wide, and toss a one-inch needle on the floor. It can then be calculated that the probability that the needle intersects a line between two strips is $2/\pi$. As I pointed out in Chapter 1, the number π may be best known as the ratio of a

[2]Relative frequencies can actually be thought of as averages. In the coin toss example, replace each H by 1 and T by 0 so that the sequence T T H T H is translated into 0 0 1 0 1. The consecutive relative frequencies are then consecutive averages of these 0s and 1s, and the expected value of one toss equals 0.5 (recall the indicators from Chapter 5).

[3]His full name was Georges-Louis Leclerc, Count of Buffon (1707-1788). He made significant contributions to biology and natural history that influenced Charles Darwin and modern ecology. To probabilists, however, he is known as a needle-tosser.

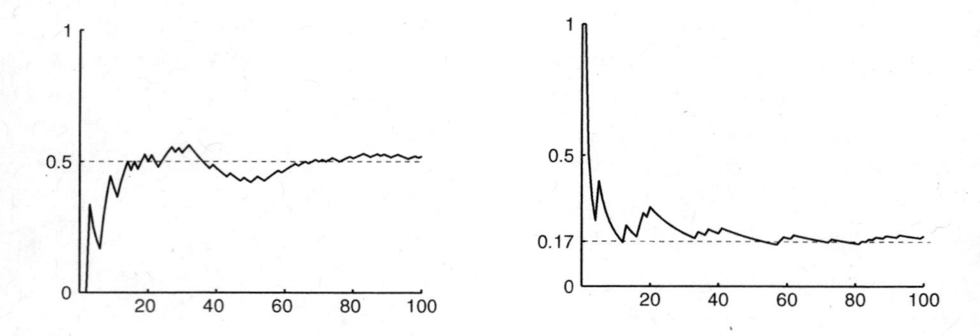

Figure 6.2 Consecutive relative frequencies of heads in 100 coin tosses (left) and of 6s in 100 die rolls (right). The success probabilities 0.5 and 0.17 are marked by dashed lines.

circle's circumference to its diameter, but it has a penchant for showing up in plenty of situations where there are no circles in sight. In this case, π does not come completely out of the blue because there are at least angles between the needle and the strips to be considered, and if you know your trigonometry, you know that π arises naturally in such contexts (if you use radians, not degrees).

Anyway, the needle tossing can be used as a method to find an approximate value of π. If you toss the needle repeatedly n times and count the number of times it crosses a line, call this number L, the law of averages tells you that

$$\text{relative frequency of crossings} = L/n \approx 2/\pi$$

from which you get $\pi \approx 2 \times n/L$. This scheme is popularly referred to as "Buffon's needle." Believe it or not but there are people who have actually done this. One of them, Italian mathematician Lazzarini, published his results in 1901, claiming to have gotten π correct to six decimal digits after 3,408 needle tosses, which is quite astonishing. It can be computed that if you toss the needle 3,408 times, your chance of getting six decimal digits correct is only about 1 in 100,000. Even getting the second decimal digit correct is no easy task; there is only a 10% chance of this. It is widely believed that Lazzarini in fact cheated. The length of his needle was $5/6$ of the width of the strips, which reduces the probability of a line crossing by a factor $5/6$, and the estimate of π then becomes $5/3 \times n/L$. Now, it is well known that π is approximated correctly to the sixth decimal by the ratio $355/113$, and Lazzarini would get this number if he got 113 crossings in 213 tosses because $5/3 \times 213/113 = 355/113$. The probability to get exactly 113 crossings in 213 is about 5.5%, and if he would fail, he could always start over and expect to succeed fairly soon. He could also continue and instead hope to get $2 \times 113 = 226$ crossings in $2 \times 213 = 426$ tosses or 339 crossings in 639 tosses and so on; any multiple of 113 and 213 will work. And, lo and behold, Lazzarini reported $16 \times 113 = 1{,}808$ crossings in $16 \times 213 = 3{,}408$ tosses! It seems very likely that he cheated but in a clever and educated way and, one might add, pretty harmless. Buffon's method may well be one of the most inefficient ways to estimate π, but it is ingenious in its connection of π to a seemingly totally unrelated random experiment.

EVEN-STEVEN? THE LAW MISUNDERSTOOD

Nobody escapes the law of averages. Fortunately. If this law breaks down, there will be anarchy of a kind that has never been witnessed before. Just picture it: You sit down at your kitchen table in the morning with your coffee and your paper and read the headline "Law of averages ruled unconstitutional!" Suddenly the cream you poured in your coffee decides to concentrate in one greasy lump on the bottom of the cup instead of being evenly mixed. The air feels thinner; almost all of the oxygen molecules have escaped to the living room. Your boss calls and tells you that you are out of work; the insurance company you work for has gone bankrupt because there was suddenly a 1,000-fold increase in car accidents. Aunt Jane calls and tells you that she has just matched the winning Powerball numbers but unfortunately, 650,000 other people had chosen the exact same numbers. When average behavior ceases to prevail, we are in trouble.[4] A short story, simply entitled *The Law*, on this theme was written by Robert Coates and published in the *New Yorker* in 1947. In this story things get so out of hand that Congress introduces legislation that requires people to be average. This also reminds me of a short story from 1964 by Swedish humorist Tage Danielsson in which Dr. Frankenstein creates Mr. Sven-Erik Average who is the average Swedish citizen, the norm for everybody. Mr. Average has 1.25 children, goes to the movies 0.75 times a week, gets half a cold every three months, and utters phrases according to how

[4]But sometimes a momentary lapse in the law can be advantageous. One evening in March 1950, all 15 members of a church choir in Beatrice, Nebraska were for various unrelated reasons more than five minutes late for choir practice scheduled to start at 7:20. At 7:25, the empty church was destroyed in an explosion. I will not comment on the possibilities of divine or diabolical intervention, but as we have seen many times, unlikely events do happen surprisingly often. The law of rare events and the law of averages coexist peacefully.

frequent they are in the language. When Mr. Average decides to revolt against the decimals that rule his life, chaos breaks out in society. Just imagine what can happen to a nation when the average citizen decides not to watch TV at night.[5]

The law of averages is different from the physical laws of nature in that it does not tell you exactly what is going to happen, only what happens on average in the long run. If you throw a coin up in the air, the law of gravity tells you that it will fall down again. The law of averages does not tell you anything. If you throw the coin up in the air hundreds of times, the law of gravity still tells you that it will fall down each time, but now the law of averages also tells you something: The coin will come down heads about 50% of the time. It is interesting that the "modern" laws of quantum physics are of a probabilistic nature, thus more similar to the law of averages than to the laws of classical physics.

The law of averages is also the reason that casinos make money. We have previously calculated that a casino keeps on average 5 cents of every $1 that is bet on roulette. What happens to one player in one spin of the wheel is highly uncertain, but when thousands and thousands of players gamble, the law of averages converts uncertainty into certainty: The gamblers may win a few battles, but the casino wins the war. In his very entertaining 2005 book *Struck by Lightning*, Canadian probabilist Jeffrey Rosenthal makes an interesting observation. When there are debates regarding government-sponsored gambling, the concern that a casino might actually lose money is never raised. Everybody knows that casinos make tons of money, and this is not from the buffet or the gift shop; it's the law of averages at work. In the movie *The Cooler*, William H. Macy plays a gloomy character who is hired by a casino to cool down "hot" gamblers simply by standing beside the game table. Hot or cold, William or no William, the law of averages reigns supreme.

Another business that relies on the law of averages is insurance. If I started an insurance company, I would calculate probabilities of different types of accidents and their associated costs and in this way determine a premium that gives me an expected gain. Let us say that I ask you to pay me $1,000 for auto insurance. I have an expected gain, but if you were my only customer I'd be in trouble if you had an accident and I had to pay. I need plenty of customers

[5]A more serious treatment of the same idea is *The Average Man* by Belgian polymath Adolphe Quetelet (1796–1874). Quetelet's most famous invention is still in used today: the *body mass index*, which is used to calculate the ideal weight of a human of a given height.

for the law of averages to kick in and guarantee a steady profit. In a way, an insurance company is like a casino where you place your bets and hope to lose. Indeed, when you buy life insurance, the insurance company predicts how long you will live and you place a bet that you will die before that.

Most people are probably familiar with some aspects of the law of averages although they may not know it in the mathematical version that we are considering. The first to formally state and prove a version of the law of averages was Swiss mathematician James Bernoulli (1654–1705), and in a letter to mathematician and philosopher Gottfried Wilhelm Leibniz, he points out that "even the stupidest man knows by some instinct of nature per se and by no previous instruction" that the law is valid.[6] Let us return once more to coin tosses. It is indeed common knowledge that if you toss a coin repeatedly, in the long run there will be equally as many heads and tails. Everything evens out eventually. However, it is important to know precisely in what sense this is true and in what sense it is not. The law of averages is frequently misunderstood and misinterpreted. What exactly does the law claim? Let us see.

It is sometimes heard that the more times you toss a coin, the more likely you are to get equally as many heads and tails. This is not just wrong; it is dead wrong! In fact, we have already seen on page 41 that the probability to get equally as many heads and tails gets *smaller and smaller* and that after $2 \times n$ tosses, the probability to have equally as many heads and tails is approximately $1/\sqrt{n \times \pi}$. But does this not contradict the law of averages? No. The notion that you would be more likely to get equally many heads and tails stems from the confusion of *absolute* and *relative* frequencies. The absolute frequency is simply the number of heads, and the relative frequency is the number of heads divided by the number of tosses. The law of averages states that the relative frequency gets close to 0.5, *not* that the absolute frequency gets close to half the number of tosses.

Consider 100 tosses. The probability to get 50 heads and 50 tails is then only about 8%. This is also the probability to get a relative frequency that is exactly equal to 0.5. The probability that there are between 45 and 55 heads

[6]James Bernoulli wrote the world's first substantial work on probability entitled *Ars Conjectandi* (the "Art of Conjecture"), published in 1713. In his honor, experiments where there are two possible outcomes (such as a coin toss) are called *Bernoulli trials*. James, also known as Jacob or Jacques, was a member of a family of prominent mathematicians including brother John and nephews Daniel and Nicholas. Mathematician E. T. Bell notes in his *Men of Mathematics* that no fewer than 120 descendants of the mathematical Bernoullis have been traced genealogically, that many of these rose to prominence in various fields, and that "none were failures."

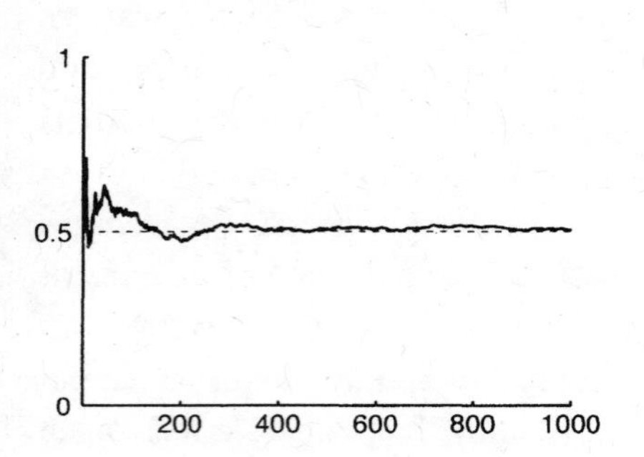

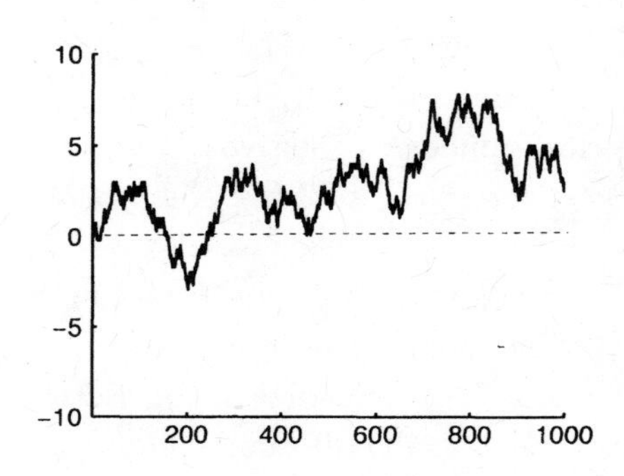

Figure 6.3 Successive relative frequencies (left) of heads and differences between the actual number of heads and the expected number of heads (right) in 1,000 coin tosses. When the graph to the right is above 0, there are more heads than tails; when it is below 0, there are more tails than heads; and at 0, there are equally as many. Note the stabilization to the left and the fluctuations to the right. Also note how the large fluctuation to the right around 100 tosses has a large impact on the relative frequency to the left but how the even larger fluctuation around 800 tosses to the right has very little impact to the left.

is about 70%, and the relative frequency is then between 0.45 and 0.55. If we instead consider 1,000 rolls, the probability to get exactly 500 heads is only about 2.5%, and the probability that the number of heads is between 495 and 505 only 25%. In contrast, the probability that the relative frequency is between 0.45 and 0.55 is now a whopping 99.8%. It gets harder and harder to keep the number of heads within a fixed distance of its expected value but easier and easier to achieve the same for the relative frequency. Figure 6.3 illustrates how the absolute frequencies fluctuate wildly at the same time as the relative frequencies quickly settle in toward 0.5.

Another, probably more common mistake, is the *gambler's fallacy*, the incorrect notion that if there are deviations in one direction, deviations in the other direction become more likely in order to compensate. Sequences containing unusually many heads are thought to be likely to be followed by sequences containing unusually many tails. Now, everybody will accept that a coin cannot remember how it came up in previous tosses, which makes such compensating behavior impossible. But a vague knowledge of the law of averages makes many believe that there still is some inexplicable compensation taking place, how else could the proportion of heads stabilize at 0.5? "Yeah, of course I understand that the coin has no memory *but*..."

But no "but..." is needed for the law to do its job. As an illustration, look at Figure 6.4, where consecutive relative frequencies in a total of 500 coin tosses are plotted. This sequence of coin tosses had an unusually large number of tails in the beginning. After 100 tosses, there had only been 35 heads, quite far from the expected number of 50, and thus the relative frequency after 100 tosses is only 0.35, as depicted in the left graph. The relative frequencies then slowly but steadily work their way up toward 0.5, and after 500 tosses, we are getting fairly close at 0.47; see the right graph. Now, this must mean that the unusually low number of heads in the first 100 tosses was compensated for by an unusually high number of heads in the last 400 tosses, right?

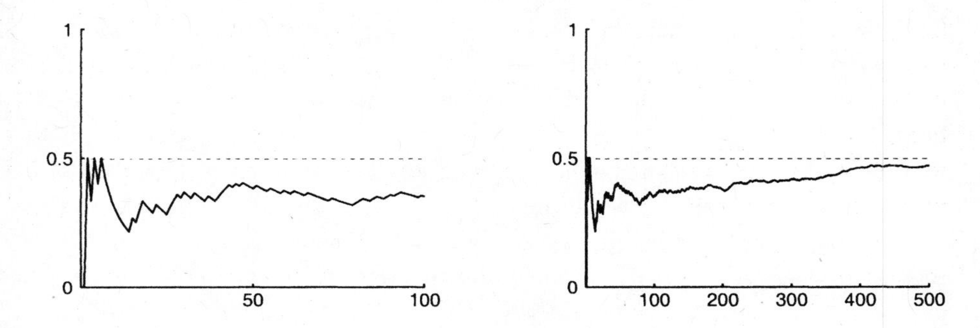

Figure 6.4 Successive relative frequencies of repeated coin tosses. The frequency of heads got a slow start in the first 100 tosses (left) but picks up gradually (right).

Wrong! Actually, there were 200 heads in the last 400 tosses, exactly the expected number. Even if we had gotten a little bit *less* than the expected number of heads, for example 180, the relative frequency would still have gone up, to $215/500 = 0.43$. Indeed, to make the relative frequency go up, we only need to beat the 35% proportion of heads from the first 100 tosses and we can certainly expect this to happen without anything mysterious occurring. As we keep tossing, the unusual number of heads in the first 100 tosses will have less and less impact on the relative frequency. Thus, there is no trend to compensate, only the persistence of average behavior that in the long run swamps the effects of any abnormalities. The stabilization of relative frequencies is not *despite* randomness, it is *because* of it.[7]

[7]My good friend, the eminent probabilist Jeff Steif, once pointed out to me over a roulette table that it is sometimes better to be completely ignorant than to have only a little bit of knowledge. After seven straight occurrences of red, the ignorant gambler does not blink (nor

When business journalist James B. Stewart wrote in his "Common Sense" column in *Smart Money*, "If a coin comes up tails 90 times out of the first 100 tosses, look for heads to make a comeback over the next 100," it set off a lively online discussion of whether he was in fact being guilty of committing the gambler's fallacy. I suppose that he by "comeback" meant that there will likely be more than 10 heads in the next 100 tosses, not more than 50 and then he was correct. We expect heads to do better than badly, not better than average. Mr. Stewart used the example to illustrate a phenomenon called *regression to the mean* ("mean" being another word for expected value). We have not mentioned our old friend Sir Francis Galton in a while, but this is the time and place: Sir Francis discovered and coined the term "regression" when he measured, among other things, heights of fathers and sons. He noticed that very tall men tended to have sons that were also taller than average, but not as tall as the fathers themselves. By inheritance, the son is expected to be taller than the average man, but if the "very tall" man is tall even by his family's standards, his son is expected to be shorter than his dad. In Galton's own words, the son of a very tall man is "probably only tall." The splendid quality of an impressive height is less pronounced in the offspring; height has regressed. Galton, in his usual politically incorrect twentieth-century style, used the term "regression to mediocrity" obviously to lament his observations that eminence tends to be less prominent in the progeny of the eminent. With a more positive outlook, he might have noticed that those who lack severely in eminence tend to get offspring that are more eminent than themselves, by the same principle. His chief interest was, however, in the upper end of the spectrum; otherwise we might have had the term "*pro*gression to the mean" instead.

For a very simple illustration of regression to the mean, suppose that you roll 6 with a die. In the next roll, do you expect to get less than 6? Yes. Do you expect to to get less than the average 3.5? No. Mr. Stewart used the phenomenon of regression to the mean to argue that oil prices, unusually high after hurricanes Katrina and Rita, were bound to come down. Whether he will be proved right or wrong, there is one problem with his analogy: the possible confusion of an extreme outcome and a change of expected value (which he also acknowledges in the article). The expected oil price is not fixed like the expected die roll is forever fixed at 3.5 and an increase in price may as well

does Jeff), whereas he who has heard of, but not understood, the law of averages believes that more occurrences of black must follow and foolishly increases his wagers.

be a repositioning of the mean as an extreme outcome (the same is of course true also for human heights but on a much slower timescale). In Chapter 8 we will study regression to the mean in more detail and take a look at what it was that Galton observed.

Speaking of the dreaded ladies K and R, after the record-breaking 2005 hurricane season was over, I heard an expert on the radio predict that the 2006 season would see fewer hurricanes than 2005. Without knowing any meteorology, I would make the same prediction. The mean number of hurricanes per year thus far is about 6, and 15 is such an extreme number that next year's number is likely to be lower by regression to the mean alone. This is true even if we are in a period of more intense hurricane activity where the mean is repositioning itself higher because 15 must still be considered an extreme outcome. If you look at how experts predict the number of hurricanes, you will notice that their predicted number is usually between the current mean and last year's number. For the upcoming 2006 season, both the two main forecasting groups NOAA (National Oceanic and Atmospheric Administration) and the hurricane team lead by Philip Klotzbach and William Gray at Colorado State University predict 9 hurricanes. These forecasts are based on a lot of climatological data and indicators; yet, in the end, it seems like Sir Francis might have come to the same conclusions. When you read this, you will know the correct answer.

COIN TOSSES AND FREEWAY CONGESTION

As simple as a sequence of coin tosses might seem, it hides a lot of surprises and unexpected behavior. To illustrate this, suppose that Tom and Harry play a game where a coin is tossed repeatedly. Harry gains a point if it shows heads and Tom gains a point if it shows tails, and as the game goes on, they keep adding up their points, a perfectly fair game. As they play on, sometimes Harry is in the lead, sometimes Tom, and occasionally they are tied. For example, if they play ten rounds and the sequence of tosses is HTHTTTHHTHH, then Harry starts out in the lead, Tom takes the lead after the fifth toss, Harry retakes the lead after seven tosses, and after ten tosses Harry is ahead by two. This is depicted in Figure 6.5. Whenever the path is above the 0 level (marked by a dashed line), there are more heads than tails and Harry is in the lead, and when the path is below the 0 level, Tom is in the lead. Positive values on the y axis (vertical axis) thus measure the size of Tom's lead, and negative values measure the size of Harry's lead. When the path hits the 0 level, they are tied. In this case, there were four ties and two changes of lead.

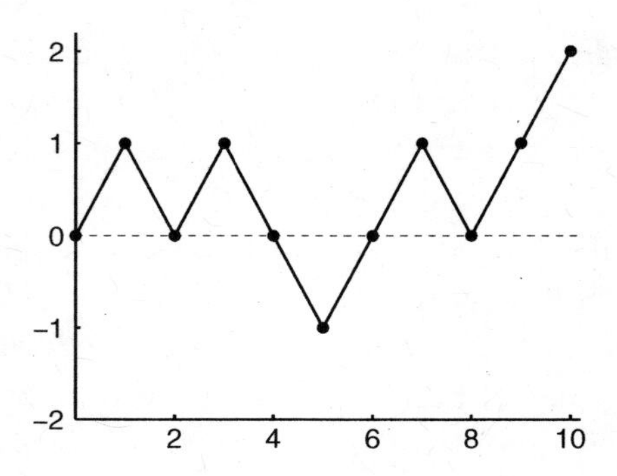

Figure 6.5 Ten rounds of Tom and Harry's coin tossing game. The dashed line is where they are tied. After ten rounds, Harry is ahead by two points.

Suppose now that Tom and Harry instead play 100 rounds and that we keep track of their scores, noting who is in the lead and when the lead changes. As the game is fair, it seems reasonable to expect each player to be in the lead about half the time, with the lead passing from one to the other every now and then. Perhaps the lead changes, what, 20 times or so?

Let me puncture these beliefs immediately. The most likely individual scenario is that the *lead never changes* and that one player is ahead all the time! There is a 15% chance that there are no changes of lead at all and the expected number of changes of lead is only about 3.5. If the lead does change, the most likely number of changes is one, followed by two, three, and so on. The probability that the lead changes more than ten times is a measly 4%. And this is true in general. Regardless of how many rounds they play, the most likely scenario is that one player stays ahead for all or most of the time. As a matter of fact, the *least* likely outcome is that each player is in the lead half the time! Of course the player ahead is equally as likely to be Tom or Harry, so on average they are in the lead equally as much, just not in the same game. Even in the simple case of ten rounds we looked at above, this becomes clear if you sit down with a pen and paper and start drawing possible paths. There are $2^{10} = 1{,}024$ of them, but you only have to draw a few in order to convince yourself that there are a lot of paths that stay entirely on one side of the 0 level and that very few paths experience frequent crossings of the 0 level.

If they instead play ten times as many rounds, 1,000, can we expect ten times as many changes of lead? No. The expected number of changes of lead in 1,000 rounds is about 12. Increase the number of games by another factor of ten to 10,000 rounds. Then only 39 changes in lead are expected. The expected number of changes of lead does not increase proportionally to the number of rounds Tom and Harry play. If they play ten times as many rounds, the expected number of changes of lead only increases by a factor of about three. It seems that changes of lead become more and more rare as the game goes on, and in a fair game, this might seem surprising. It shouldn't. Remember how we have seen that ties become increasingly improbable and that after $2 \times n$ tosses, the probability of a tie is about $1/\sqrt{\pi \times n}$. Thus, after 100 tosses, the probability of a tie is about 8% (this is $n = 50$); after 1,000 tosses ($n = 500$), about 2.5%; and after 10,000 tosses ($n = 5{,}000$), the chance of a tie is only 0.8%. But this also means that changes of lead occur more and more seldom because a change of lead must be preceded by a tie. Indeed, each time there is a tie, there is a 50–50 chance that it is followed by a change of lead; it happens if the player who was behind before the tie is successful in the following toss.

In long sequences we don't expect many ties, which means that we instead expect long excursions either above or below the 0 level. Occasionally there is a tie and there is then a 50% chance of another tie two steps later. The path might stay near the 0 level for a while, experience a few ties and changes of lead, but eventually it will stray again away from the 0 level. The clumping phenomenon that we saw examples of in Chapter 3 shows up again. Ties tend to appear near each other in clumps with long spaces in between. Thus, ties are not only rare, but they are also highly irregular. Once the path strays away from the 0 level, it just keeps finding its random way one step at a time and there is is no invisible force that works to pull it back toward 0, so once it gets far away, it will take it a very long time to come back.

Figure 6.6 shows some computer simulations of 1,000 rounds of the coin tossing game. That is, three of them do and the fourth is fake. It is almost impossible to get a simulation outcome that looks like the upper left graph. I had to cheat (big time) to get that scenario, simulating little pieces at a time, making sure the path started returning toward 0 once it tended to get away. And that's precisely why this path is by far the least typical of the four; what I did is not done by randomness. The other three are obtained from computer simulations of sequences of independent coin tosses and are in no way extreme. In fact, I even weeded out several simulation runs that gave

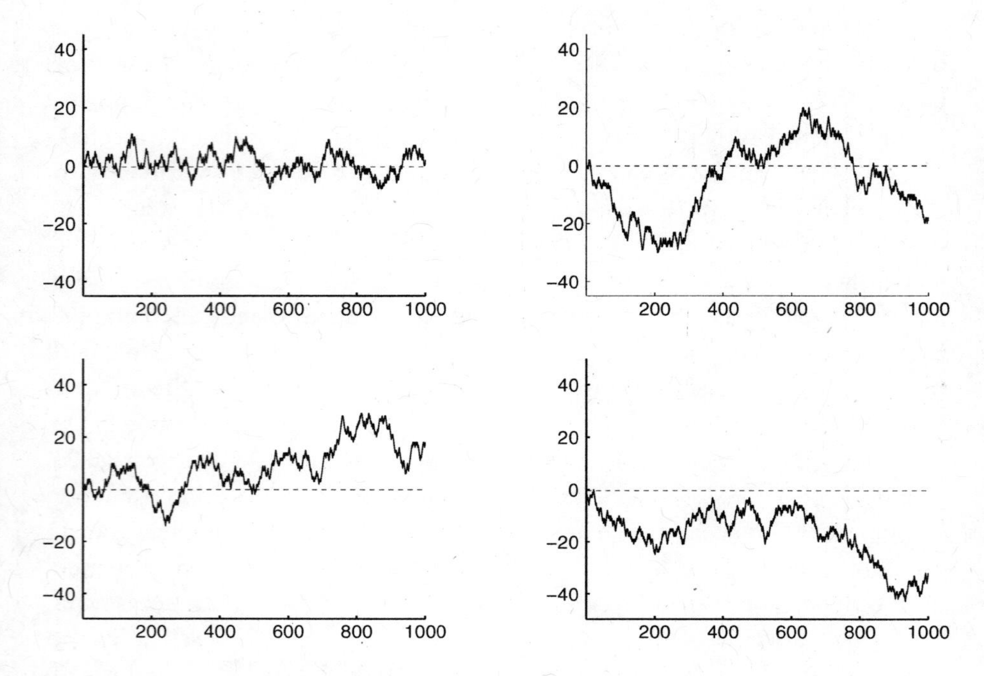

Figure 6.6 Possible outcomes of 1,000 rounds of the coin tossing game. The upper left is fake and very unlikely to occur at random; the other three are typical computer simulation outcomes. Note how the ties tend to occur in clumps.

much more extreme outcomes. Note the long excursions and few changes of lead in the three real graphs, and note how ties tend to occur in clumps.

If presented with the graphs in Figure 6.6, it would not surprise me if most people would pick out the one in the upper left corner as the truly random one. In fact, it would surprise me if they didn't. Isn't it interesting how we are constantly accompanied by randomness in our journey through life, be it on a deeper level by quantum physics or genetics, or by the multitude of mundane risks and chances we experience on a daily basis; yet, we are so inept at recognizing true randomness when faced with it?

Is it possible that the path leaves the 0 level never to return again? No, and we can easily prove it with a recursive use of the law of total probability, very much similar to how we solved the racket sports problems in Chapter 1 and the branching process problem of extinction of a cell population on page 65. Let us walk the path starting at 0. The first step has to be either up to 1 or down to -1. Let us suppose that it is up. We are then at 1, and the question becomes about what the probability is that we will eventually return to 0 again. Call this unknown probability p. Next take the second step. With probability 0.5, we are back at 0 again, and with probability 0.5, we are at 2, two steps away from 0. Here comes the neat part. In order to eventually get back to 0 from 2, we must first get back to 1, and this has probability p because getting from 2 to 1 is just as likely as getting from 1 to 0. And once back at 1, we must then get back from 1 to 0, and this has probability p so the probability to get back from 2 to 0 is simply $p \times p = p^2$ (we are using independence here). But this means that p satisfies the equation

$$p = 0.5 \times 1 + 0.5 \times p^2$$

where the "1" is the (obvious) probability that we make it back to 0 if the step from 1 is downward. As you can see by plugging in $p = 1$, this solves the equation and the eventual return to 0 is certain. (If you are familiar with quadratic equations, you know that there can be more than one solution, but in this case there isn't; 1 is a "double root.") Finally, recall that this was the case when the very first step was upward from 0 to 1, but clearly the calculations are the same if the step is downward, replacing 1 and 2 by -1 and -2. We have proved that the path will eventually return to 0, and when it is there, it starts a new journey eventually to return to 0 again and so on, forever and ever.

There are many situations in real life that resemble the coin tossing example. In *Taking Chances*, John Haigh draws a parallel with sitting in rush hour traffic. You and the car next to you are equally as likely to advance one car length, and by a coin toss, St. Christopher, Patron Saint of Travelers, decides who it will be. If you lose the toss, you'd better catch up quickly or you will be left behind. Chances are good that you will watch your competitor slowly disappear in the distance, one car length at a time. And should you manage to catch up and leave him behind, you quickly forget him and focus on the fresh competitor at your side. You seldom feel like a winner in rush hour traffic.

If two equally good sports teams played each other repeatedly, we would expect to see something similar to the coin tossing graphs. Of course there are lots of factors influencing a sports game and it would be impossible to account for them all, but nevertheless, let us look at two famous long running competitions: the Army–Navy football game and the Oxford–Cambridge boat race.

The Army–Navy game has been played annually since 1890 (with ten of the years canceled), and after the 2005 game, Navy leads 50–49. There have been 7 ties. The boat race between the two giants in British higher education

Oxford University and the University of Cambridge has been held annually since 1829 except for during the two world wars and a few more cancellations. After the 2005 race, Cambridge is in the lead 78–72. There has been one dead heat, in 1877. The sequences of games in both the football game and the boat race are depicted in Figure 6.7, and it is remarkable how the typical features of coin toss sequences are present (ties and canceled games have been excluded, so the match between year and score is not perfect). Army has been ahead far more often than Navy and had a period when they were ahead for 48 consecutive years. As for the boat race, the teams have been in the lead about the same amount of time, but notice how seldom the lead has changed. Right now in 2005, Cambridge has been in the lead for 70 years.

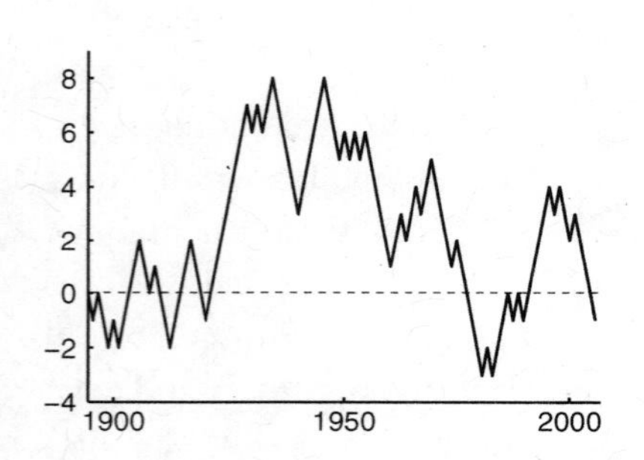

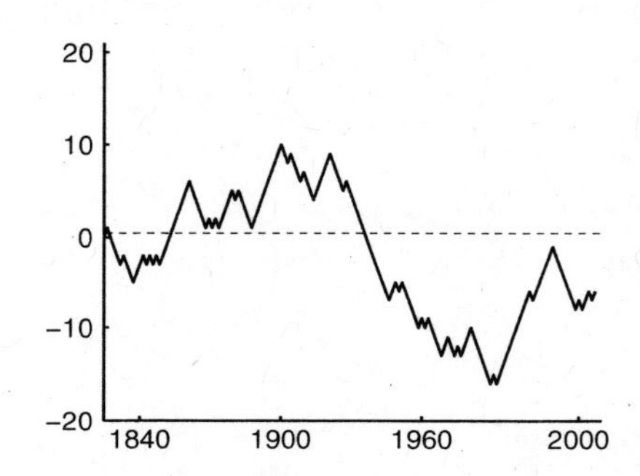

Figure 6.7 The Army–Navy football game since 1890 (left) and the Oxford–Cambridge boat race since 1829 (right). Army and Oxford are ahead above the 0 level. Years with ties or canceled games are not included.

Less famous but equally fiercely competitive is the annual track and field meet between Sweden and Finland, in Swedish called *Finnkampen* (the "Finn Battle") and in Finnish *Suomi-Ruotsi-maaottelu* ("Finland–Sweden International"). This event consists of all Olympic track and field events, with three competitors from each country in each event (there is traditionally a lot of elbowing in the middle-distance running, and in 1992 all six runners in the 1,500-m race were disqualified). The men's battle has been going on since 1925 with a break during World War II, and the women's battle first took place in 1953. Figure 6.8 shows the sequences of wins and losses for the men's and women's battles. Again, typical coin tossing patterns emerge. In 2005, the Swedish men are far behind at 40–25 but the Swedish women lead 27–23. Incidentally, when my sports interest was at an all time high during the 1970s, Sweden did not win a single battle, men or women, between 1972 and 1979. Tough times. But knowing what I know about random path behavior, I rest assured that the Swedish men will eventually get in the lead (my Finnish friends somehow fail to accept this inevitable fact).

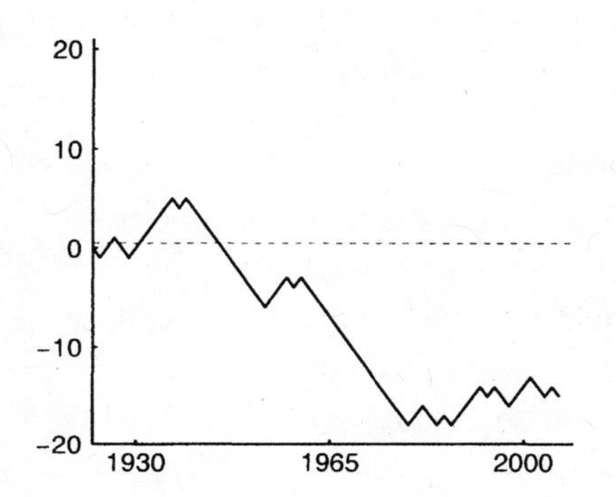

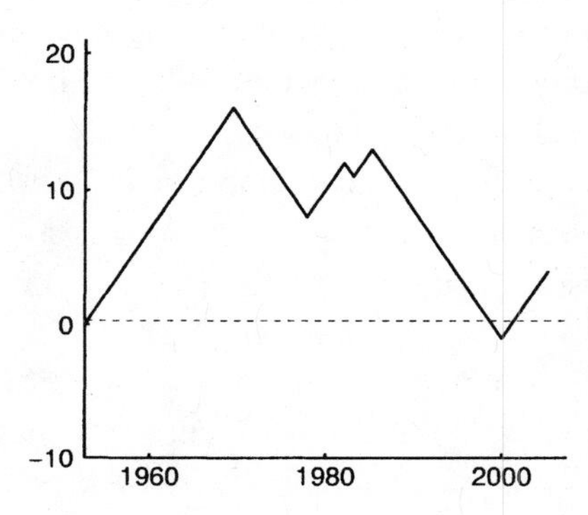

Figure 6.8 The "Finn battle" since 1925 for men (left) and since 1953 for women (right). Sweden is ahead above the 0 level.

The coin tossing graphs are examples of what probabilists call *random walks*. In popular presentations, they are often illustrated by a drunkard who is staggering back and forth, each time equally as likely to take a step to the left as to the right or a person lost in a blizzard (preferably sober) who is staggering in the same way. As the coin toss is fair, the walk is equally as likely to go up as it is to go down in each step, and we call such a walk *symmetric*. For a random walk that is not symmetric, place repeated odd bets in roulette. Your chance of winning is then $18/38 \approx 0.47$ in each round, slightly smaller than 0.5, so if you graph your winnings, the resulting walk is asymmetric, each time a little more likely to go down than up. For short sequences it will be hard to tell the difference between this and repeated coin tosses, but in the long run, there will be no doubt when the roulette graph eventually stays on the negative side forever. In this case, unlike the coin tosses, there is an invisible force that is constantly working to pull the graph downward, and the law of averages tells us that, although this force may experience some minor setbacks, it will be successful in the long run. It can be computed that you can expect to break even (visit the 0 level) 19 times before you finally set foot on the inevitable path to ruin.

LET'S GET SERIOUS

This section is devoted to the mathematical formulation and proof of the law of averages. Even if you fear theory and formulas, I would encourage you not to skip this section just yet. It may be a bit more challenging than previous sections but gives a nice little glimpse of what the world of advanced probability theory looks like. Still, if you wish you can safely skip this section and proceed to the next.

Before we continue, let us recall the notion of a random variable from Chapter 5. A random variable is something that gets its numerical value from an experiment. For example, if you roll a die, you can describe the outcome by the random variable X, which can take on the values 1 through 6. Suppose instead that you intend to roll the die ten times. You will then get ten outcomes, and before the experiment you can describe this by the ten random variables $X_1, X_2, ..., X_{10}$. In one way these are all the same because they describe the same experiment and therefore have the same probabilities of the same outcomes. We say that they are *identically distributed*. But in another way they are not at all the same because each of them takes on a value that can be anything from 1 to 6 regardless of the values of the others. We say that they are *independent* (recall independent events from Chapter 1).

In the same way we can describe many situations of repeated observations or measurements as sequences of independent and identically distributed random variables, but let us stick with the dice for a while. Remember now that the law is about averages. The average of the n first random variables is denoted by $\bar{X}$ ("X bar"), in a formula

$$\bar{X} = \frac{X_1 + X_2 + \cdots + X_n}{n}$$

and the law of averages says that $\bar{X}$ approaches 3.5 as n increases or, in other words, that $\bar{X}$ will be arbitrarily close to 3.5 if only n is large enough. This is a little problematic though. For example, when can we be certain that the average is between 3.4 and 3.6? Perhaps after 1,000 rolls? 10,000? But how could we possibly guarantee that there, for example, isn't an unusually large number of 5s and 6s to follow that will bring a later average above 3.6? Well, we can't. We can never say for *certain* that from now on will the average stay between 3.4 and 3.6 on but have to be satisfied with a conclusion that states that this becomes *more and more probable* as the number of rolls increases. For example, if you roll the die n times and consider the average

$\bar{X}$, how large must n be so that you can be at least 99% certain that $\bar{X}$ is between 3.4 and 3.6? In a formula, we are looking for the n that is such that $P(3.4 \leq \bar{X} \leq 3.6) \geq 0.99$, or equivalently, the n that is such that

$$P(|\bar{X} - 3.5| \leq 0.1) \geq 0.99$$

This problem is difficult to solve exactly. Let us first examine $\bar{X}$ a little closer. What can we say about its expected value and variance? As $\bar{X}$ is computed from the random variables $X_1, ..., X_n$, it is a random variable. Thus, $\bar{X}$ has an expected value and a variance, and to find them, note that $\bar{X}$ is obtained by adding $X_1, ..., X_n$ and then dividing by n, and we can use the properties of expected values and variances from Chapter 5. First, the expected value of the sum $X_1 + X_2 + \cdots + X_n$ is $3.5 + 3.5 + \cdots + 3.5 = n \times 3.5$. Second, since $\bar{X}$ is the sum divided by n, by linearity the same is true for the expected value and thus the expected value of $\bar{X}$ is $n \times 3.5/n = 3.5$. The average of n die rolls has the same expected value as a single die roll, not so strange if you think about it. From the previous chapter, we also know that the variance of a single die roll is 2.9, and by additivity of variances, the sum $X_1 + X_2 + \cdots + X_n$ has variance $n \times 2.9$.[8] Finally, to get the variance of $\bar{X}$, remember that variances are not linear but that constants are squared. In this case, as $\bar{X}$ is the sum divided by n, the variance of $\bar{X}$ is the variance of the sum divided by n^2, that is, $2.9/n$. To summarize, the expected value of the average of n die rolls is 3.5 and the variance is $2.9/n$. Note how the variance decreases with increasing n, which indicates that there is less and less variability in $\bar{X}$ as the number of rolls increases, indicating that $\bar{X}$ tends to be close to 3.5.

So far so good, but we want to be more specific. This is where we must apologize to Mr. Chebyshev for making a wee bit of fun of his inequality in Chapter 5 because now it turns out to be really useful. Remember that Chebyshev taught us that the probability that a random variable is within k standard deviations of its expected value μ is at least $1 - 1/k^2$. In our case, the random variable of interest is $\bar{X}$, which has expected value 3.5 and variance $2.9/n$. The standard deviation is therefore $\sqrt{2.9/n} \approx 1.7/\sqrt{n}$, and Chebyshev's in-

[8]The additivity of variances was mentioned in footnote 6 on page 139. The "restrictions" that are there referred to are that the variables are independent. If they are not, variances are not necessarily additive. For example, if X is the outcome of a die roll and we double it to $2 \times X$, this doubled outcome has a variance that is the variance of X quadrupled to $4 \times 2.9 = 11.6$. As $2 \times X$ can also be written $X + X$ and as this does *not* have variance $2.9 + 2.9 = 5.8$, the variance is not additive in this case (and clearly X cannot be independent of itself).

equality applied to these numbers gives, in a formula,

$$P(|\bar{X} - 3.5| \leq k \times 1.7/\sqrt{n}) \geq 1 - 1/k^2$$

which looks similar to the inequality that we are interested in:

$$P(|\bar{X} - 3.5| \leq 0.1) \geq 0.99$$

Remember now that n is the unknown number that we are after. Identifying the right-hand sides of the two expressions above informs us that we should set k equal to 10 (as $1 - 1/100 = 0.99$). Identifying the left-hand sides informs us that we should set $k \times 1.7/\sqrt{n}$ equal to 0.1. As $k = 10$, we get the equation $17/\sqrt{n} = 0.1$, which finally gives $\sqrt{n} = 170$ and hence $n = 170^2 = 28{,}900$. Let us round it up to 29,000 and summarize what we have just shown: If you roll a die *at least* 29,000 times, you can be *at least* 99% certain that the average is between 3.4 and 3.6. Now, that is a lot of rolls, but we did not do this for practical purposes. Rather, it is an example that illustrates that we can always find an n that achieves any specified level of certainty (less than 100%) that the average is within any specified distance of 3.5. Our particular choices were 99% certainty and distance 0.1, but the principle clearly works for any numbers. For example, similar calculations show that in order to be at least 99.9% sure that the average is between 3.45 and 3.55, we need at least 1,160,000 rolls.

Let us attack the general case. We now have a sequence $X_1, X_2, \ldots$ of random variables that are independent and identically distributed. In particular, they all have the same expected value μ and variance σ^2, and the average $\bar{X}$ therefore has expected value μ and variance σ^2/n, that is, standard deviation $\sigma/\sqrt{n}$. We can redo the calculation above to conclude that in order to be at least 99% certain that the average $\bar{X}$ is between $\mu - 0.1$ and $\mu + 0.1$, n must be at least $10{,}000 \times \sigma^2$. This is more general than than the die rolls because it is true for any μ and σ^2, and note how the value of n is affected by the value of σ^2. If the variance σ^2 is large, we need to add more variables to subdue the variability in $\bar{X}$.

Now the final touch. In a mathematical proof, we don't fiddle around with different values of the distance from μ and the level of certainty and see what happens. Rather, we regard the distance and level of certainty as fixed but arbitrary constants and give them variable names. First, let us denote the distance from μ by ϵ (Greek letter "epsilon"). Second, let us denote the level of certainty by $1 - \delta$ (Greek letter "delta"). We are interested in small distances and large probabilities (i.e., small δ), and it is a mathematical tradition to use the letters ϵ and δ to denote numbers that are really small. Thus, we want the probability that $\bar{X}$ is between $\mu - \epsilon$ and $\mu + \epsilon$ to be at least $1 - \delta$. (We are doing math now and then; probabilities are always numbers between 0 and 1, not percentages, but if you insist on the latter, you can always say $100 \times (1 - \delta)\%$ instead.) In a formula, we want n to be such that

$$P(|\bar{X} - \mu| \leq \epsilon) \geq 1 - \delta$$

and because Chebyshev's inequality applied to $\bar{X}$ tells us that

$$P(|\bar{X} - \mu| \leq k \times \sigma/\sqrt{n}) \geq 1 - 1/k^2$$

we can again identify the right-hand sides to conclude that we should take $k = 1/\sqrt{\delta}$ (remember that δ is fixed to start with). Next, use this value of k and identify the left-hand sides to get the equation $\epsilon = \sigma/\sqrt{n \times \delta}$. Solve this for n to get $n \geq \sigma^2/(\delta \times \epsilon^2)$ (the "$\geq$" is there because n must be an integer). In the special case of die rolls that we started out with, we had $\sigma = 1.7$, $\epsilon = 0.1$, and $\delta = 0.01$ (check this for yourself).

We are done. We have managed to show that, for any sequence of random variables and any ϵ and δ, regardless of how ridiculously small they are, we can always make sure that $\bar{X}$ is between $\mu - \epsilon$ and $\mu + \epsilon$ with a probability that is at least $1 - \delta$ if only n is large enough. You want to be 95% certain that $\bar{X}$ is between $\mu - 0.5$ and $\mu + 0.5$? Easy. This means that your ϵ is 0.5 and your δ is 0.05, and because $1/(0.05 \times 0.5^2) = 80$, you just need to let n be at least $80 \times \sigma^2$. Perhaps 99% certain? No problem, just increase n to be at least $400 \times \sigma^2$. At least $100 \times (1 - \delta)\%$ certain? Most certainly, just make sure that your n is at least $4 \times \sigma^2/\delta$.

You may be familiar with the concept of *convergence* of sequences. A sequence of real numbers $x_1, x_1, ...$ is said to converge to a number x if the x_n eventually come arbitrarily close to x. Formally, for any ϵ, the absolute value $|x_n - x|$ will be less than ϵ if only n is large enough. We say that x_n *converges to* x as n *goes to infinity* and write $x_n \to x$ as $n \to \infty$ (and the number x is called the *limit* of the sequence). For example, the sequence $\{1/n, n = 1, 2, ...\}$ (that is, the sequence of numbers $1, 1/2, 1/3, ...$) converges to 0 because however small ϵ is, if n is larger than $1/\epsilon$, then $1/n$ is smaller than ϵ (as are all subsequent numbers in the sequence). For another example, on page 87, I mentioned that the sequence $(1 - 1/n)^n$ converges to e^{-1}. This convergence can be proved strictly with ϵ and δ, but this is not the place to do it. Now consider our sequence of consecutive averages. This approaches μ but in a different way from how $1/n$ approaches 0 or $(1 - 1/n)^n$ approaches e^{-1}. Although we can never be certain that the average is within distance ϵ of μ, the law of averages tells us that we can be certain that it is so with an arbitrarily *high probability* by choosing n large enough. We still say that $\bar{X}$ converges to μ but add the little qualifier *in probability*. Thus, if you roll a die repeatedly, the sequence of successive averages converges in probability to 3.5, and if you toss a coin repeatedly, the sequence of successive relative frequencies of heads converges in probability to 0.5. As we are dealing with random phenomena, this is about as good as it gets.[9]

In footnote 2 on page 147, I pointed out that relative frequencies can be interpreted as averages and probabilities as expected values, which means that we have also shown that successive relative frequencies of an event that each time occurs with probability p converge in probability to p. The stabilization of relative frequencies is therefore merely a special case of the law of averages, but historically it was the first result of this type to be proved, done by James Bernoulli whom I mentioned earlier. His proof is much more complicated than ours because Chebyshev's inequality was of course not known to him in the early eighteenth century. More general versions of the law of averages were proved during the twentieth century, mainly by Russian mathematicians and most notably the late, great Andrey Kolmogorov, founder of modern probability theory who also appeared in an anecdote about independence in Chapter 1.

BELLS AND BREAD

You have probably seen it before: the bell-shaped curve that describes datasets as diverse as widths of tree trunks, weights of bags of nuts, stock market fluctuations, light intensity, errors in astronomical measurements, and IQ scores.

[9]Actually, it does get a little better and it is possible to show that there is what probabilists call *convergence almost surely*. This means that there will always be an n such that $\bar{X}$ stays within ϵ of μ forever after; we just can't say in advance what this n is. The "almost" part refers to the fact that we can still imagine a sequence that does not converge to 3.5, for example, an infinite sequence of 6s (it's easy if you try), but this will simply never happen (it has probability 0). The topic of different *modes of convergence* is a profound and fascinating part of probability theory, but this is not the time or place to treat it in any detail.

The curve is known as the *normal distribution* or the *Gaussian distribution*, both names being slightly unfortunate. There is nothing abnormal about data that do not have a normal distribution, and Carl-Friedrich Gauss (1777–1855) who is known as the "prince of mathematics" (there is no king) may be the greatest of them all, but he did not discover the bell curve. Not that he had much of a chance as the first description of the bell curve was published over half a century before Gauss was born, by French mathematician Abraham de Moivre (1667–1754) in his 1718 book *The Doctrine of Chances*. Our friend Sir Francis Galton was among the first to suggest the name "normal distribution," and that is the term that is most commonly used by us probabilists and statisticians, whereas engineers and scientists seem to prefer "Gaussian." Then again, what's in a name? That which we from now on will call a normal distribution by any other name would look the same. Like this:

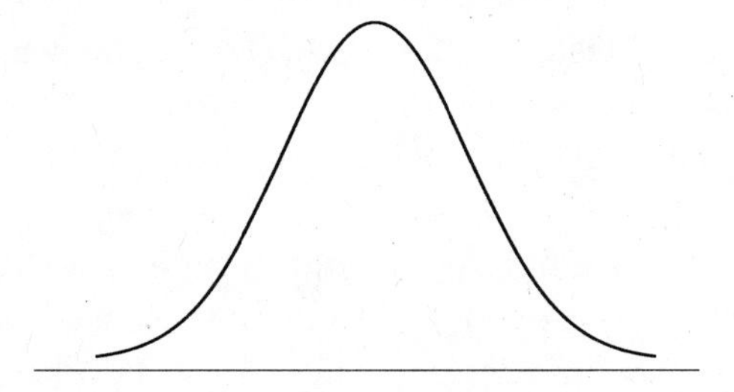

Figure 6.9 The bell curve of the normal distribution. Pure, simple, serene.

Although Gauss did not discover the bell curve, he made extensive use of it, noticing that measurement errors, for example, in astronomical observations tended to be distributed according to the bell curve. Before the switch to the euro, the German ten mark bill had a picture of Gauss, a picture of the bell curve, and even stated the mathematical equation that describes the bell curve. The equation is not quite as attractive as the curve itself:

$$f(x) = \frac{1}{\sigma\sqrt{2\pi}} e^{-(x-\mu)^2/2\sigma^2}$$

but before you disregard it, note a few things. There is a μ in there that is the expected value located at the center of the bell. There is also a σ which is the standard deviation. It can be shown that for a normal distribution, the probabilities that an observation falls within one, two, and three standard deviations of μ are about 68%, 95%, and 99.7%, respectively. To compute probabilities from the bell curve, we need to consider the area under the curve. As μ is in the middle of the symmetric bell, the probability that an observation is above μ is 50% because the area under half the curve accounts for 50% of the entire area. The area under the curve between $\mu - \sigma$ and $\mu + \sigma$ accounts for 68% and so on. Thus, for data from a normal distribution, very few observations will be outside the interval $\mu \pm 3 \times \sigma$. Even fewer observations, only two in a billion, will be outside $\mu \pm 6 \times \sigma$. If you manufacture something that has a normal distribution and get an observation outside six σ of μ, you have either seen something extremely unlikely or there is something wrong with your manufacturing process. You'd better look it over. This approach is an example of *statistical quality control*, which has been used extensively and saved companies a lot of money in the last couple of decades. The term *Six Sigma*, a registered trademark of Motorola, has evolved to denote a methodology to monitor, control, and improve products and processes. There are Six Sigma societies, institutes, and conferences, and you can even get a Six Sigma Black Belt (but might have to settle for the Green Belt). Whatever Six Sigma has grown into, it all started with considerations regarding the normal distribution.

Finally, before we leave the equation for the bell curve to its fate, note that it contains the number π. Again, no circles and no needles to toss either. It just pops up. If only Archimedes could be with us, how pleased he would be.

So what does it then mean to have a normal distribution? It means that if you take repeated measurements and plot them in a graph, the graph will resemble the bell curve above. Now, the bell curve is nice and smooth and the graph of actual observations never is. However, the key word is "resemble." We will not see the exact bell curve, but we will clearly see the bell shape being indicated, particularly if we have many measurements. See Figure 6.10 for two datasets with indicated bell curves. One is a classic, 100 measurements of the speed of light ($\times$1,000 km/second), conducted by Albert Abraham Michelson in 1879. The other is contemporary, heights in inches of the 67 players on the Chicago Bears roster for the 2006 season. Note that in order to get anything that can be compared with the bell curve, we need to group the measurements into classes. The closer the class is to the center of the bell, the more measurements it will contain. The two sets are different from each other in the sense that the speed of light is constant, so here the normal distribution arises from measurement error, whereas in the football player data, the normal distribution describes individual variation around the team mean height.

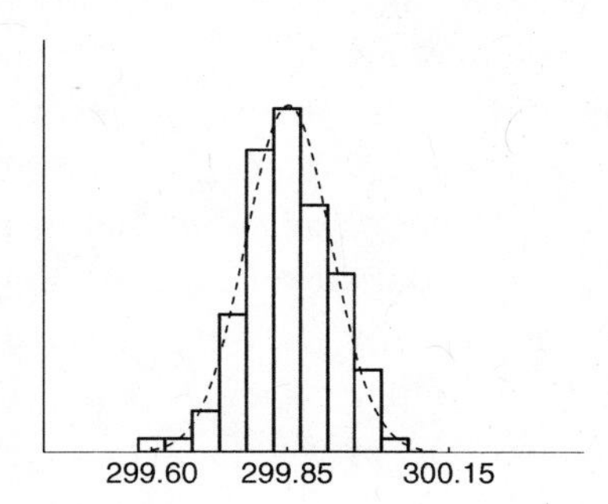

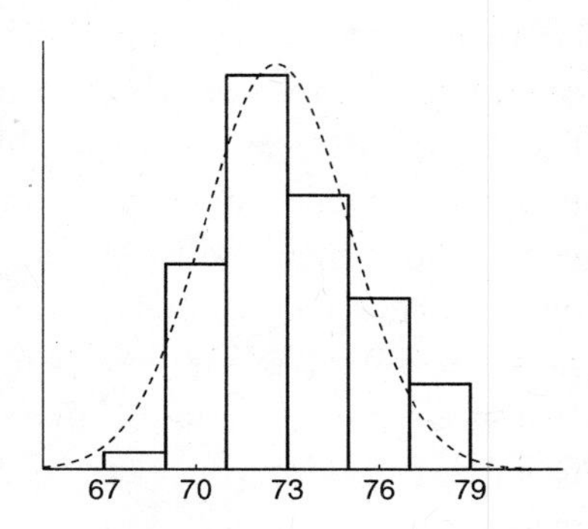

Figure 6.10 Two datasets with the corresponding bell curves dashed. To the left are 100 measurements of the speed of light in $\times$ 1,000 km/s (grouped into ten speed classes) and to the right are 67 heights of Chicago Bears players in inches (grouped into six height classes).

A somewhat controversial quantity that is often assumed to follow a normal distribution is IQ score where the tests are calibrated to make the expected value 100 and the standard deviation 15. This means that half of us are above 100 and half below, 68% of us are between 85 and 115, and 95% of us between 70 and 130. Only 2.5% have an IQ score above 130 (as the area between 70 and 130 accounts for 95%, the remaining 5% are equally divided below 70 and above 130). Recall Marilyn vos Savant and her record-breaking score of 228, which is 8.5 standard deviations away from the expected value. The probability of getting an IQ score higher than Marilyn's is so extremely small that she can consider her *Parade* column safe for the rest of her life. Incidentally, "grading on a curve" is also based on the normal distribution, with the assumption that test scores follow the bell curve, which I as a teacher don't think should be followed too slavishly. I remember how a classmate in high school was once told that he could not get the highest grade because that had already been awarded to two other students so there was none left for him. Clearly senseless, but my friend did not suffer any severe psychological damage and is now a prominent neuroscientist who also knows his mathematics.

There is an anecdote about the French mathematician Henri Poincaré (1854–1912) that goes something like this: Poincaré bought loaves of bread from a certain bakery. The loaves were supposed to weigh a kilogram each, but Poincaré methodically weighed every loaf, plotted his measurements, and got what looked like a normal distribution with expected value 950 grams. He complained to the police who told the baker to stop cheating. Over the next year, Poincaré's loaves weighed on average a kilogram, but he nevertheless claimed that the baker still cheated. The reason for this was that Poincaré's data, while averaging a kilogram, did not look like a normal distribution. Rather, his observations looked like a normal distribution with expected value 950 grams but cut off to give an average of 1,000 grams. Poincaré concluded that the loaves still followed the same normal distribution as before but that the dishonest baker instead of giving Poincaré a randomly selected loaf set aside one of his heavier loaves for the mathematician, still cheating the other customers. See Figure 6.11 for an illustration with computer-simulated values of what it could have looked like. To the left are 100 loaves with an average weight of 950 grams and to the right, 100 of the heavier loaves that the baker set aside, averaging 1,000 grams. Poincaré was able to detect this by not just considering the average weight but also the shape of the distribution of weights. Although the anecdote illustrates a clever way to use probabilities to detect cheating, I doubt that it is true. What Frenchman who has just bought a loaf of bread would go home and weigh it? I'd like to think that he would be more concerned with what cheese and wine to buy to go with the bread than how much it weighs. Besides, don't you think the local baker and the local policeman would have known each other?

Why did Poincaré's loaves follow a normal distribution? And why would the bags of nuts that I mentioned above do it? Why is it that the bell shape

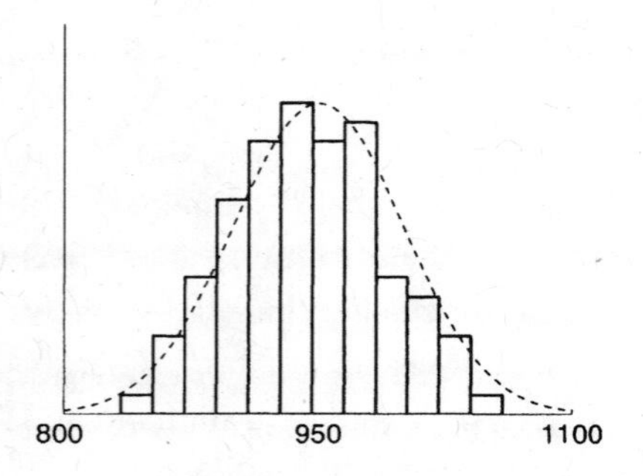

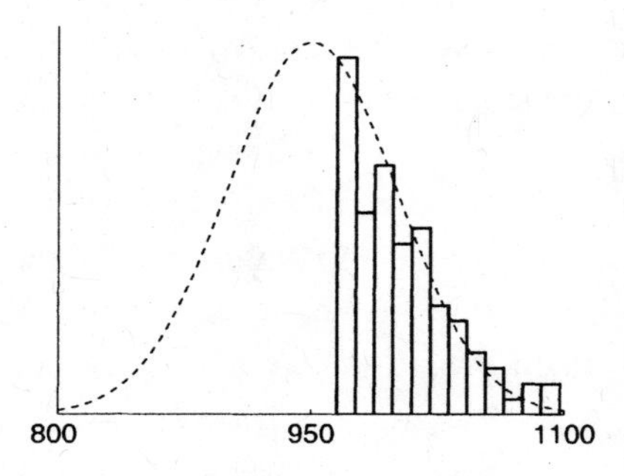

Figure 6.11 Poincaré's loaves. To the left are 100 loaves with an average of 950 grams and to the right are the loaves that the baker selected for Poincaré, averaging a kilogram but still baked according to the same normal distribution with an average weight of 950 grams.

shows up in so many situations? The answer lies in a remarkable theoretical result in probability that we will investigate in the next section.

HOW A TORONTO QUINCUNX CHANGED MY LIFE

The section title immediately invites two questions: (1) What is a quincunx? and (2) How can anything in Toronto change anybody's life anyway? Let me immediately in response to (2) point out that I think Toronto is a wonderful city with great Portuguese restaurants, excellent public transportation, and of course the Maple Leafs. For our purposes, question number one is of more interest though. A quincunx is a device made out of nails placed on a board so that they form a large triangle: One nail at the very top, two underneath, three next, and so on on so forth. A plate of glass is placed over the nails and the contraption is put in an upright position. Literally, quincunx refers to a pattern of five dots arranged like they are on a die with one dot in the middle and the remaining four in each corner. Our quincunx gadget is in fact made up of many little quincunxes. Other names are *bean machine* or *Galton box*, and indeed, it was invented by Sir Francis to illustrate the normal distribution. But how?

Start by dropping a metal ball on the top nail. The ball will then bounce either to the left or to the right, and after that it hits another nail with the same two possibilities, go left or right. It bounces like this until it finally hits the bottom where a row of containers is placed. The leftmost and rightmost containers are hardest to hit because they require that all bounces are in the same direction; there is only one path to each of them. The containers in the middle are easiest to hit because there are many different paths that lead to them. As you keep dropping balls, you will see them arrange themselves nicely according to the bell curve over the containers, most in the middle, fewer to the left and right, and fewest at the very ends.

The quincunx gives an illustration of an important result in probability called the *central limit theorem.* This theorem states that if you sum many independent random variables, the sum tends to have a normal distribution regardless of what are the original random variables. In everyday language, this means that if you measure something that is composed of many small independent contributions, it will likely follow a normal distribution regardless of what are the contributions themselves.[10] Poincaré's loaves were certainly composed of many little particles of flour, salt, and yeast, and this explains the normal distribution that he observed. And the same goes for the bags of nuts. The weight is the sum of the weights of many individual nuts, and even if they are different, for example, a mixture of small hazelnuts and larger Brazil nuts, the resulting sum still follows a normal distribution. The central limit theorem thus explains the universal presence of the normal distribution. As soon as a quantity can be described in terms of many small independent contributions, the bell curve enters. For the quincunx, the final position of a ball is the sum of the contributions from each step to the left or right.

So what about my life-changing experience in Toronto then? It's not too exciting a story. I mentioned my year off from college when a friend and I traveled around the world. Other than the more exotic stops such as Samoa, Fiji, and Tonga, we also spent a few days in Toronto, and it was there, at a science museum, that I first saw a quincunx in action. The bell-shaped curve was painted on the glass, and as I heard a father explain the mechanism to his son, I decided that I would study probability instead of computer science upon my return to Sweden. That's all.

The proof of the central limit theorem is complicated and uses methods that are far beyond what I can describe in this book. I find it absolutely fascinating that such complicated theory explains such easily observable phenomena as Poincaré's bread and Galton's quincunx. The central limit theorem also gives more insight into the coin tossing games we investigated and graphed earlier. If Tom and Harry play 1,000 rounds, the expected score is that they are tied at the 0 level in the graph because the game is symmetric but we have seen that it is not a very likely outcome. The standard deviation of the position after 1,000 rounds can be shown to equal 33, and by the central limit theorem, the position has approximately a normal distribution. Moreover, by what I mentioned about the normal distribution above, about 68% of observations are within one standard deviation of the expected value, and for the coin tossing game, this means that 32%, about one third of the games, end with one

[10]Such contributions are said to be *additive*. There are also many situations where contributions instead are *multiplicative*, for example, stock prices that tend to change in proportion to their value. If you are familiar with logarithms, you know that the logarithm of a product equals the sum of the logarithms and the logarithmic stock market data are therefore more likely to follow a normal distribution than the data themselves. If you go to any financial webpage, you will notice that you can choose to get the charts of stocks, funds, and indexes in either linear or logarithmic scale.

player being ahead by at least 33 points. In general, the standard deviation after n rounds is $\sqrt{n}$, so if 10,000 rounds are played, there is a one in three chance that either Tom or Harry is up at least \$100.

The central limit theorem also supplements the law of averages to give additional information about the average $\bar{X}$. As we have seen, the law of averages says that $\bar{X}$ tends to be near μ and the central limit theorem tells us that the distance between $\bar{X}$ and μ has approximately a normal distribution with expected value 0 and standard deviation $\sigma/\sqrt{n}$. This can be used to compute probabilities involving $\bar{X}$, and because a relative frequency is a special kind of average, the theorem applies to relative frequencies as well. When I computed the probabilities in connection with Buffon's needle problem above, that is precisely what I did.

FINAL WORD

The law of averages and the central limit theorem are the two most fundamental results in probability. You could perhaps say that the law of averages is no surprise; it would certainly be strange if it was not true. But it serves as certification that the theory of probability has been developed in a logical manner, consistent with the real world, and that our interpretations of expected values and probabilities as long-term averages and relative frequencies are valid (perhaps that's why we call it a "law" and not a theorem). The central limit theorem is more flashy with a touch of magic. Whenever we study anything that is composed of many small independent contributions, the bell curve with its complicated mathematical equation shows up by necessity whether we like it or not (but we do). The law of averages and the central limit theorem work in tandem to describe what happens to averages and relative frequencies in the long run, and together they bring law, order, and simplicity into a chaotic and complicated world.

7

Gambling Probabilities: Why Donald Trump Is Richer than You

FRENCH LETTERS

Considering how many ways people have invented to gamble away their money, this chapter could easily be the longest in the book. Gambling authority John Scarne's 1986 *New Complete Guide to Gambling* covers all the casino games, card games, dog and horse racing, slot machines, and plenty more, including several games he invented himself. The book presents a lot of odds calculations, but it is not a probability book. Still, it is almost 900 pages long. For a splendid account on probability and gambling more along the lines of the book you are now reading, see the previously mentioned *Winning with Probability* by John Haigh. His book covers not only the traditional casino games but also board games such as Monopoly and Backgammon and TV game shows such as *Weakest Link* and *Who Wants to Be a Millionaire*.

I will be more modest and only cover a few of the most common casino games. With this intention, this chapter could instead easily be the shortest: By the law of averages the casino wins and you lose. Period.

Still, gambling is interesting for several reasons. First, of course, because people gamble, the law of averages notwithstanding. And they gamble a lot; gambling is a multi-billion dollar industry in the United States alone and a vital source of income for states, Indian tribes, charitable organizations, and Donald Trump. And then there is the casino in Monte Carlo, the Royal Ascot horse race, the casinos in Macau, the football (soccer) pools in Europe, the

Nenana Ice Classic where residents of Alaska bet on the month, day, hour, and minute the Tanana River ice will break, and so on and so forth. And then there's the Internet...

Second, gambling is particularly interesting to probabilists because it reveals randomness in a very pure form. Probability calculations are much easier and more accurate for lotteries and roulette than for complicated phenomena like the weather or the stock market. The history of probability is also intimately connected with gambling. I mentioned in Chapter 1 how Galileo helped gambling Florentine noblemen solve a problem involving dice, but it was another noble gambler who finally initiated the systematic study of probability: Frenchman Antoine Gombaud (1607–1648) better known as Chevalier de Méré.

De Méré knew that if he bet on getting at least one 6 in four rolls of a die, he would have a slight edge. We have actually already computed the probability of this event in Chapter 1, but since it's been a while, let us do it again just for practice. As "at least one 6" is the opposite of "no 6s," this is a case for Trick Number One and we get

$$\text{P(at least one 6)} = 1 - \text{P(no 6s)} = 1 - (5/6)^4 \approx 0.52$$

so de Méré would have a 52% chance to win this bet. Another popular game was to roll a pair of dice and look out for a *sonnez*, a double six. If you bet on at least one double six, how many times do you need to roll a pair of dice in order to make the bet favorable? An old gambling rule said that as there are six times as many outcomes of two dice as of one, you need to roll six times as many. Thus, getting at least one double six in 24 rolls would give a favorable bet. However, something caught de Méré's suspicion and he contacted his friend Blaise Pascal (1623–1662), a brilliant mathematician and scientist, who among other things built a mechanical calculator, invented a famous triangle, and lent his name to the metric unit for pressure. Pascal in turn wrote a letter to another brilliant French mathematician, Pierre de Fermat (1601–1655), who gave name to the world's most famous mathematical theorem. The reason for its notoriety has changed though; *Fermat's last theorem* used to be famous as a one-line theorem that nobody could prove; now it is famous as a one-line theorem with a 200-page proof. At any rate, Pascal's letter to Fermat set off a correspondence between the two that is usually considered the origin of probability theory. To us who master Trick Number One to perfection, the solution of de Méré's problem is of course simple: The probability to get a double six in one roll of the pair is 1/36, and thus, the probability to get at least one double six in 24 rolls is

$$\text{P(at least one double 6)} = 1 - (35/36)^{24} \approx 0.49$$

so this bet would not be in de Méré's favor.

It is unclear if de Méré's suspicion was brought on by persistent gambling losses as is often claimed or by his intellectual abilities, of which Pascal testified in a letter to Fermat, calling the Chevalier "very able" although "not a geometrician," which Pascal considered a "great defect." Some claim that a success probability of 49% is such a small disadvantage that it would have been impossible for de Méré to discover empirically that the game was unfavorable. I disagree and offer two arguments, one utilitarian and one mathematical. The utilitarian argument is that in this game, out of every 100 *pistoles*, the Chevalier bet, the house kept 51, and it gave back 49. Thus, the house gained on average 2 pistoles for every 100 that was bet, a gain of 2%. The casino in Monte Carlo only keeps 1.35% on even-money bets in roulette (bets that pay as much as you wagered such as bets on red/black or even/odd), and they do not do it for charitable reasons. Also, we will later see that there are bets in craps where the house only keep about 1.4%. I don't know on how large a scale gambling was done in the Chevalier's days. It was certainly a pleasure reserved for the privileged, but an expected house profit of 2% could potentially be enough to make a consistent profit and the Chevalier and other gamblers would indeed pay for it.

The mathematical argument is more convincing and goes as follows. If you start with a small initial fortune and play until you have either doubled your fortune or gone broke, it is possible to compute the probability that you go broke first. This depends on your success probability, that is, the probability that you win in a single round. Let us say that you start with five pistoles and roll the pair of dice. Your success probability, the probability to get at least one 6, is 49%, and it can be computed that the probability that you go broke before doubling your fortune is about 54%. If De Méré really believed the old gambling rule, he would instead expect a success probability of 52%, and then the probability to go broke before doubling is only about 41%. If the Chevalier gambled vigorously, starting over with 5 pistoles fortunes again and again, he would likely be able to tell the difference between a 54% and a 41% "broke rate." If he instead started with 10 pistoles, the corresponding rates would be 59% and 33%, quite a difference. And if he started with 100 pistoles, a success probability of 52% makes it virtually certain that he doubles before going broke. The probability of this is 99.97%; only once every 3,000 games would he go broke before doubling his fortune. In contrast, with the true success probability of 49%, there is a 97% chance to go broke before doubling. I'd say if this happened just once, it would be enough to make anybody suspicious. I will later demonstrate how these calculations were done. The Chevalier would of course have to gamble a lot to be able to verify this; it can be shown that if he starts with 100 pistoles, the expected duration of the game until he either goes broke or reaches 200 pistoles is about 2,500 rounds. Still, the dice roll fast and as he had time to spare and no boss to report to, why not?

It is interesting to note how a very small house advantage in each round adds up to an enormous advantage in the long run. This observation also serves to illustrate an important principle for those who want to run a casino. A game must be favorable to the house but not too favorable or people will not gamble. And even with a very small edge per round, the house advantage over time is merciless. In the Chevalier's time, with the old gambling rule, the two-dice game even seemed to be disadvantageous to the house and gamblers might think that an oversight had been made. Even in our day and age when we all know that casinos take our money, the expected loss in some games is still so small that we have a good chance of winning a bit and might even occasionally lose our mathematical bearings and believe that the house can actually be beaten. It can't.

Chevalier de Méré also suggested another much more intricate problem to Pascal, the *problem of points*. Let Tom and Harry play their coin tossing game where heads gives Harry a point and tails gives Tom a point. Suppose that they wager $50 each and that whoever is the first to get six points wins the money. After nine tosses, Harry leads 5–4 and the atmosphere is tense. The coin is tossed and...falls down the gutter drain. They have no more coins and need to decide how to divide the money. Tom suggests that as they didn't finish the game, each should just get his $50 back. Harry, on the other hand, thinks that he should get all the money because he was in the lead when the game was interrupted. Tom counters that because Harry's lead was only 5–4, they should then at least divide the money in proportions five to four, which would give Harry $56 and Tom $44.

Neither of these solutions appealed to Pascal and Fermat. In their letters they instead agreed on an ingenious solution: The stake should be divided *in proportion to the probabilities* that each player would win if they had been able to finish the game. As Tom can only win by tossing two tails in a row, the probability that he wins is $1/4$ and Harry wins with probability $3/4$. As Harry is three times as likely to win as Tom, he should get three times as much money, which gives him $75 and Tom $25. This is very reasonable. If they were to start from score 5–4 and play first to six over and over, out of every four games, Tom would win on average one and Harry three and thus Harry would end up with three times as much money. The brilliant Frenchmen's solution is the only way to divide the money that is entirely consistent with our notion of expected value and historically the first example of sophisticated probabilistic reasoning. Any compulsively gambling French nobleman could attempt to calculate odds on various bets, but the problem of points required some really innovative ways of thinking by the sharpest brains around. In Pascal's own words:

> I am so overjoyed with our agreement. I see that the truth is the same in Toulouse as in Paris.
>
> Pascal in a letter to Fermat, July 1654

As a sidenote, it is interesting that it took so long for mathematics and games of chance to meet. The ancient Greeks who were outstanding mathematicians also very much enjoyed to roll the dice but never attempted to describe games of chance in mathematical terms. Perhaps the Greek mathematicians were too purist to dabble with something as impure as randomness and something as worldly as dice. Perhaps they thought that mathematics could only deal with the absolute, the unchangeable, but whatever the reason, the emergence of a theory of probability had to wait until the Renaissance. It is amusing that the theory of probability that nowadays has so many practical applications started with a bunch of idle noblemen killing time by rolling dice.

ROULETTE: A CLASSY WAY TO WASTE YOUR MONEY

It is sometimes claimed that roulette was invented by Pascal, and sometimes it is said to have been invented by a French monk in order to break the monastic monotony (if this is true, I suppose that he didn't notice that the sum of the roulette numbers equals 666). It has also been claimed that the origins of roulette can be traced to an old Chinese game, but there does not seem to be much evidence supporting that theory. More certain is that roulette as we know it spread across Europe and America during the nineteenth century and that the differences between the "French wheel" and "American wheel" were then established. Both wheels have the numbers 1–36 and 0, but the American wheel also has 00, a double zero, which increases the house advantage (and as far as I have been able to find out, the American wheel is the older of the two and the double zero was at some point dropped in Europe). We have already computed that the house keeps 5% (more accurately 5.26%) in American roulette. The corresponding house profit in European roulette is 2.7%. The payouts in European roulette are the same as in American roulette, but the odds are slightly better due to the lack of a double zero. For example, if you bet $1 on odd you win $1 with probability 18/37 and lose it with probability 19/37, which gives an expected gain of

$$1 \times 18/37 + (-1) \times 19/37 = -1/37 \approx -0.027$$

an expected house profit of 2.7%. The expected percentage that the house gains is called the *house edge*. On even-money bets, the house edge is simply the difference between the probability that you win and the probability that you lose. To see this, let the probability that you win be p and the probability that you lose $q = 1 - p$. As games are stacked against you, $p < q$ and your expected gain is

$$1 \times p + (-1) \times q = p - q$$

which makes the house edge $q - p$. For example, if you have a 48% chance of winning, out of every $100 you wager, you lose $52 to the house, win $48 and keep the $48 that you wagered in your winning bets. That means that you have a total of $48 + $48 = $96 and the house has gained $4 out of $100 or 4%, which is also the difference between your winning probability 52% and your losing probability 48%.

Some casinos, most notably the one in Monte Carlo, have an additional rule about even-money bets being "imprisoned" when 0 comes up, giving the player a second chance to win. This further reduces the house edge to 1.35%. The price you have to pay for this is that you must learn not to bet until the croupier says *Faites vos jeux* and stop when he calls out *Rien ne va plus*; European roulette is played in French. If this sounds too challenging, Atlantic City roulette has instead the redeeming feature of only taking half the money on even-money bets, giving the house a 2.6% edge. Betting $1 in European roulette, by the way, shouldn't it be euros, pounds, or Swiss francs? Not necessarily, you can play European roulette in America too. Tom Ainslee reports in his 1987 book *How to Gamble in a Casino* that the

European roulette tables in Nevada are often completely empty, whereas the less favorable American tables across the isle are packed with players who obviously enjoy doubling their expected losses. Let the poet speak:

> In play there are two pleasures for your choosing –
> The one is winning, and the other losing.
>
> Lord Byron, *Don Juan*, 1823

I have previously pointed out that there are many different bets you can place in roulette but that regardless of how you bet, the payouts are calculated so that you will have your expected 5 cents loss per $1 regardless of how you bet. Let us see how the payouts are calculated. The basic bet is the straight bet where you bet on a single number. The probability that you win is $1/38$, and on page 139, we calculated the expected gain on a straight bet to be $-2/38$. Now suppose that you instead place a *split bet*, which means that you place your chip on the border between two numbers that are adjacent on the table. You win if any of these numbers come up, so you now have a probability of $2/38$ to win. How much? Let us denote the payout by a, compute your expected gain, and set it equal to $-2/38$ to see what a must be. This gives the equation

$$a \times 2/38 + (-1) \times 36/38 = -2/38$$

where we get rid of the 38s and move 36 to the right-hand side and we have

$$2 \times a = 34$$

so the payout is $17 if you bet $1, or as it is commonly written 17:1. There are other bets. In a *street bet* you place your chip so that it marks a row of three numbers, in a *square bet* you place it so that it marks four numbers, and you already know about odd/even bets and red/black bets. There are some more bets, but they all have in common that you place your wager on a block of numbers and if any of these come up, you win. Let us find a general formula for the payout, again denoted by a. It your bet covers n different numbers, the probability that you win $\$a$ is $n/38$ and the probability that you lose $1 is $(38-n)/38$. Your expected gain set equal to $-2/38$ now gives the equation

$$a \times n/38 + (-1) \times (38 - n)/38 = -2/38$$

which has the solution $a = 36/n - 1$ and hence

$$\text{payout in a } n\text{-number bet} = (36/n - 1){:}1$$

A street bet bet is $n = 3$ and the formula gives the payout 11:1, and a corner bet is $n = 4$ and the payout is 8:1. In a *column bet* you bet on the 12 numbers in any of the three columns on the table, and with $n = 12$, the payout is 2:1. You see now why a roulette table has 36 and not 35 or 37 slots; as 36 is divisible by all numbers 2, 3, 4, 6, 9, 12, and 18, it is easy to calculate the exact payouts that give the house the same edge on any bet.

No rule is without exception though. In a bet called the "five number line," you mark the five numbers 00, 0, 1, 2, and 3. In our formula, this is $n = 5$ and we get a payout of $36/5 - 1 = 6.2$, so the correct payout is 6.2:1. But casinos don't like to cut their chips in fifths and only offer 6:1 on this bet. As the probability of winning is $5/38$, your expected gain with the five-number line bet is

$$6 \times 5/38 + (-1) \times 33/38 = -3/38 \approx -0.08$$

an expected loss of about 8 cents per \$1 instead of 5 cents. In a way this means that a good betting strategy actually does exist in roulette: Avoid the five-number line! Following this piece of advice will cut your losses by over 30% and I offer it for free.

Not everybody is as benevolent as yours truly. There are people out there who devise systems to beat the roulette odds. They are either shameless hucksters trying to sell you something or, well, really out there. One procedure, sometimes known as the "Black System" has been promoted as a miraculous way to beat the odds. The system goes like this: You bet \$1 on the middle column and \$2 on black. The middle column pays 2:1 and black pays 1:1, even money. Now, the middle column contains eight black numbers, so if any of these comes up, you win \$4 (\$2 on black and \$2 on the middle column). If any of the other ten black numbers come up, you win \$1 (win \$2 on black, lose \$1 on the middle column). That is 18 numbers out of 38 where you win. But you also break even in four other cases: If you get any of the four red numbers in the middle column you lose \$2 on black and win \$2 in the middle column. Thus, you win in 18 cases, break even in 4 cases, leaving only 16 cases for the house to win! I'm sure this might sound convincing if sold by some smooth-talking hustler to slightly inebriated mathematically challenged gamblers but you and I quickly find the catch: In the 16 cases in which you lose, you lose \$3 and it is easy to see that the usual expected 5% loss prevails. Of course we already knew this from our knowledge of expected values, but these are not arguments the smooth talker is willing to buy. Academic nonsense, no, he knows by *personal experience* that his system works.

The most famous betting system in roulette is the *martingale*.[1] It is also the one that is hardest to refute because it really seems to work. The basic and simple idea is to bet on black (or any other even-money bet). Start betting \$1. If you win, you are ahead and can set your profit aside and play again. If you lose, you increase your next bet to \$2. If you lose again you bet \$4, next time \$8, and so on, each time doubling your bets. Now, red or zeros will not keep showing up forever and by the time black finally comes up, you win one dollar more than your accumulated losses. For example, if you get three occurrences of red followed by black you have lost $\$1 + \$2 + \$4 = \$7$ and then bet \$8 to win \$8 in the fourth round. Try some other values and check that this always works. Eureka...?

Not quite. There are two problems, one practical and one fundamental. The practical problem is that casinos have limits on how much you can bet. As the consecutive doublings add up quickly, you may hit the ceiling after not too many rounds. If the table minimum is \$10 and the limit \$500, you would be out of luck after six unsuccessful bets and would not be allowed to place the seventh bet of \$640. The probability of six consecutive unsuccessful bets is $(20/38)^6 \approx 0.02$, which is small but in no way negligible and you have then lost \$630.

The fundamental problem also has to do with the fact that successive doublings add up quickly but is much more severe: Even without table limits, in order to make the martingale system work, you must be infinitely wealthy! This may sound absurd, but it can be shown that even if the game was fair, your expected loss would actually be infinite. The assumption of a fair game makes the expected value a little easier to compute. Let us see how much you can expect to lose before you win. If your first win comes immediately,

[1] In advanced probability theory, a martingale (a term of equestrian origin) is a process whose fundamental property is that it stays on average the same. The symmetric random walks we saw in Chapter 6 are examples. As they are equally as likely to step up or down, on average they stay where they are (the expected value of the change equals 0). Now, roulette does not give a symmetric random walk so the process is not actually a martingale but something called a *supermartingale* (it's super for the casino, not for you). The name martingale refers to the particular betting scheme described here.

you have lost nothing and this has probability $1/2$. If it comes in the second round, you have first lost \$1 and this has probability $1/2 \times 1/2 = 1/4$. If it comes in the third round, you have first lost $\$1 + \$2 = \$3$ dollars and this has probability $1/8$ and so on. Generally, if you lose k times before winning, it can be shown that you have lost a total of $2^k - 1$ and that this has probability $1/2^{k+1}$. As there is no upper limit on the number of rounds you must wait, we get

$$\text{expected loss} = \sum_{k=1}^{\infty} (2^k - 1) \times 1/2^{k+1}$$

and it is not too hard to show that this sum is actually infinite, meaning that as you keep adding terms, the sum just keeps getting bigger and bigger without bounds. The "-1" eventually becomes negligible next to the huge number 2^k, and as this is multiplied by $1/2^{k+1}$, we get $1/2$. Thus, $1/2$ keeps being added to the sum forever and ever and there is no end to how large the sum becomes.[2] This seems a bit counterintuitive. As your win must finally come, you can never *actually* lose an infinite amount of money, but as we have seen before, the expected value does not have to be a possible actual value. The practical consequence of the infinite expected loss is simply that regardless of how much money you have, you may play the martingale successfully for a while, perhaps a long while, but eventually you are bound to hit a losing streak that is long enough to ruin you. The martingale system is not unique to roulette and can be used in any game, with any odds, but unless the game is to your advantage to start with, it does not work. Then again, should you find a game that is to your advantage, there is no need for a strategy. Just sit back, relax, and watch the law of averages make you rich.

CRAPS: NOT SO DICEY AFTER ALL

Craps is roulette's noisy neighbor. When all chips have been placed on the roulette table, there is that moment of silent anticipation when all you can hear is the ball whirling slower and slower, then bouncing around in the pockets of the wheel until it finally decides who of the gamblers gets to cheer and

[2]The infinite expected value in this game was discussed by cousins Nicholas and Daniel Bernoulli of the famous Bernoullis mentioned in Chapter 6. It is usually referred to as the *St.Petersburg paradox*, and in order to resolve the problem, Daniel Bernoulli introduced the concept of expected *utility*, which became a central idea in economics.

who gets to sigh. Nothing like it in craps. The dice roll fast, there is constant yelling and cheering; it is a fast and lively game. There are some different versions of the history of the game, but it seems to have developed from the older similar game *hazard* (from Arabic *az-zahr* meaning "the die"), which was played in France and England. The stories how it became established in America also differ, but because I currently live in Louisiana, I will go with the one that claims that the game was brought by the Acadians from Nova Scotia and that "craps" is the Cajun word for crabs, which in turn was the nickname for the lowest roll.

Craps is played with two dice whose sum is computed. There is a slew of different bets that can be placed, and unlike in roulette, these are not equally as unfavorable but range from a house edge of 1.4% to 16.7% (there are also so-called *odds bets* that are completely fair, but in order to place one, you must first have placed another bet). Let us focus on one of the most favorable and thus most popular bets, the *pass line bet*. This is an even-money bet, which means that when you win, you win as much as you wagered. The dice are rolled, and if the sum is 7 or 11 (a *natural*), you win. If the sum is 2, 3, or 12 (*craps*), you lose. Any other sum establishes the *point*, and the dice are rolled until either the point is repeated (in which case you win) or 7 shows up (in which case you lose). What is the probability that you win?

You are by now familiar with how to compute probabilities for two dice. Let us consider the first roll. Out of 36 possible outcomes, 6 (1–6, 2–5, ..., 6–1) give the sum 7 and 2 (5–6 and 6–5) give the sum 11, giving you a probability of $8/36$ or about 22% to win already in the first roll. As there are only four ways in which you can lose in the first roll (1–1, 1–2, 2–1, 6–6), the probability of this is only $4/36$ or about 11%. With probability 67%, neither of this happens and the probability that you win then depends on what your point is. Let us say that you roll a three and a five in the first roll, thus establishing the point 8. The dice are now rolled until either 7 or 8 shows up. What is the probability that 8 comes first so that you win? As there are six ways to get sum 7 and five ways to get sum 8, out of the 11 cases in which either 7 or 8 shows up, 5 makes you the winner and 6 the loser. Thus, if your point is 8, the probability that you win is $5/11$ or about 45%. Together with 6, 8 is the best point you can have followed by 5 or 9, the worst points being 4 or 10. Let us say that you are unfortunate enough to roll 4. You must then succeed to repeat this before 7 shows up, and as your three cases (1–3, 2–2, 3–1) are competing with the six cases to get 7, your chances of winning are now $3/9$, about 33%. In the intermediate case when your point is 5 or 9, you have a 40% chance of winning.

If it wasn't for the fact that craps is actually offered by casinos, it wouldn't be immediately obvious that it is unfavorable. In 22% of the rounds, you win immediately, twice as often as you lose immediately. In the remaining 67% of the rounds, you have at worst a 33% chance to win but it is more likely that you do better because the better points are more likely than the worse. To compute the probability that you win, we can use the law of total probability applied to four cases in which you have a chance of winning. See Table 7.1 for these cases, their probabilities, and the probability of winning in each case. The probabilities are expressed as fractions so that you can easily check for yourself that they are correct.

Table 7.1 Initial roll in craps and associated probabilities

First roll	7 or 11	6 or 8	5 or 9	4 or 10	2, 3, or 12
Probability	$8/36$	$10/36$	$8/36$	$6/36$	$4/36$
P(win)	1	$5/11$	$4/10$	$3/9$	0

Finally, multiply the probability of each case with the probability of winning in each case:

$$8/36 + 10/36 \times 5/11 + 8/36 \times 4/10 + 6/36 \times 3/9 \approx 0.493$$

that is, an overall winning probability of about 49.3%. Wow, that is so close to 50%! Maybe there is an error somewhere and the game is actually fair? No, a careful calculation reveals that the probability of winning is exactly $244/495$, which equals 0.492929... and the game of craps is indeed stacked against you. Not by much though, and you will not find many casino games that give you better odds. If you play 1,000 rounds at $1 a round, you win on average 493 times, thus raking in $986, making the house edge only 1.4%. However, the game is also the fastest in the casino so you might well end up going broke at the craps table before your friends at the roulette table with the 5% house edge.

Even more intricate than the pass line bet is the *don't pass line bet*, which is almost the exact opposite of the pass line bet. You lose if if the first roll gives 7 or 11, win if it gives 2 or 3, and if the game goes on you win on 7 and lose on the point. In essence, this means that you are betting that the casino will win, a bet that sounds too good to be true. And it is. Note how 12 is missing from the craps numbers above and there is the rub. Although the house wins on 12 on a pass line bet, you don't win on 12 on the don't pass line bet. A sum of 12 in the first roll is considered a tie and you get your money back but don't win or lose anything. This little caveat may not sound like such a big deal but is precisely what is needed to turn the law of averages against you and, instead of letting you get the house's previous edge of 1.4%, the house still gets an edge, this time of 1.37%. Who told you that life was fair? Craps is an excellent example of how a game is created to make the house advantage very small but still large enough for the law of averages to work for the casino on a daily basis.

I mentioned that not all craps bets are equally as unfavorable. The worst bet you can place is to bet on 7. It is as simple as it sounds; you just place your chip where it says "Seven" on the table and win if the first roll gives 7. The payout is 4 to 1 and as the probability to get 7 is $6/36 = 1/6$, your expected gain is

$$4 \times 1/6 + (-1) \times 5/6 = -1/6 \approx -0.167$$

which is a house edge of 16.7%! The only reason to play this is if 7 has not shown up in a very long time so that you know that it is due to show up soon. No, wait, that didn't sound right. You have to wait until 7 is hot and shows up frequently. No, wait,...

BLACKJACK: MONEY FOR MNEMONICS

The origin of blackjack is as unclear as that of roulette or craps. It seems to have made its way into the casinos a little later than the other two, in the early 1900s but had existed as a private card game known as "twenty-one" or "vingt-et-un" long before. The basic rules are simple. You are dealt cards one by one by a dealer and add their value to compute your total. Jacks, queens, and kings count as 10 (from now on, when I say 10 this means any of the cards that count as 10) and aces count as 1 or 11. Your objective is to beat the dealer who will deal his own cards after all the players. However, if you get more than 21 you go bust and lose your wager. This means that after each card, you need to decide whether to *hit* (get another card) or to *stand* (get no more cards). Thus, there is a fundamental difference between blackjack and most other casino games because you can influence your chances of winning during the game. When you have decided to stand, the dealer gets his cards and has to keep hitting until he has at least 17 and then he has to stand. The dealer's total is compared with yours and whoever has the highest wins (even money for you). If you have the same total, it's a tie (or a *push*) and no money changes hands, and if the dealer goes bust, you win (even money). There is one exception: If your first two cards are an ace and a 10, you have a "blackjack" and win 1.5 times your wager immediately. If the dealer gets a blackjack, he takes your wager immediately (with some exceptions). If you both get blackjacks, it is a tie. Some special rules give you more options such as *doubling down*, which gives you the option of doubling your wager if you think that you have a favorable hand after the first two cards, and *splitting*, which gives you the option to split up a pair into two separate hands if you think it is to your advantage.

This game seems very fair. Whoever has the highest total wins, and if there is a tie, nobody wins. In fact, it may even seem favorable to you because you can choose your strategy, whereas the dealer has no choice whatsoever. Moreover, you get paid more for a blackjack than the house does and you have those extra perks of being able to double down or split if you want to. What is the catch then? The asymmetry in going bust: If you go bust, you lose your wager directly and the dealer does not even have to play. This unfairness of the game more than compensates the house for your advantages. Blackjack is still among the least unfair of the casino games, and if you like Dustin Hoffman's Raymond in *Rain Man* can memorize cards, you can even turn the game into your favor. Actually, you don't have to be quite that good; there are card-counting systems in which you keep track of a total score by adding one for high cards, subtracting one for low cards, and keeping the score unchanged for intermediate cards. You then adjust your strategy and the size of your bets according to what is your total score. Needless to say, this requires a lot of practice and it will still give a very slim edge, 2% at the very best. Also remember the fate of Raymond and his brother Charlie in Rain Man; casinos don't like card counters. And they will not send a William H. Macy-type cooler to the table; they will throw you out.

The probability calculations involved in blackjack are substantially more complicated than those of roulette or craps. The proportions of different cards in the deck change as the game goes on. For example, if the game is played with one deck and three aces have been dealt, you know that there is only one more ace left. To eliminate possible advantages for the player by such considerations, blackjack is usually played with at least six decks of cards and they are reshuffled long before they run out. For simplicity, let us assume that the proportions of the cards stay the same throughout the game so that all the cards 2–ace are always equally as likely. As 10–king count as 10, there is then a $4/13$ or about 31% probability to get a 10 and a $1/13$ or about 7.7% probability of any other number, 1–9 and 11 (you can choose whether to count ace as 1 or 11, and you can change your mind as often as you want). The probability that you win of course also depends on how you play. Suppose that you follow the exact same strategy as the dealer: Hit on 16 and lower, stand on 17 and higher. The possible results are then 17, 18, 19, 20, 21, "blackjack," and "bust," where 21 means that the result is 21 but not a blackjack, for example 7–8–6. The probabilities of the possible results are listed in Table 7.2. They are rounded to integer values; in reality, 17 is slightly more likely than 18 and a blackjack is slightly less likely than 5%.

Table 7.2 Final results and their probabilities for the dealers's strategy

Final result	17	18	19	20	21	Blackjack	Bust
Probability	15%	15%	14%	18%	5%	5%	28%

Let us now see what happens if you wager $10 and both you and the dealer play according to the dealer's strategy. Let us disregard splitting, doubling, and all other special rules but keep your extra payout for a blackjack. You then have one advantage and one disadvantage. Your advantage is that you get paid $15 if you get a blackjack and the dealer does not. The probability that you get a blackjack is 5%, and the probability that the dealer does not get one is 95%, so the probability that you win $15 on a blackjack is $0.05 \times 0.95 = 0.048$ or 4.8%. You get to cash in on your advantage less than once every 20 hands. In contrast, your disadvantage to go bust, in which case, the dealer wins without even having to play happens with probability 28%, more than once every four hands. If you don't go bust or get a blackjack, the probability that you win depends on what is your result. For example, if you have 17, you can only win if the dealer goes bust, which has probability 28%. However, if the dealer does not go bust, you don't necessarily lose; if the dealer also gets 17, it is a tie and this has probability 15%. You lose if the dealer gets 18–21 or blackjack, which has probability 57%. If your result is 18, you win if the dealer has 17 or goes bust, which has probability 43%, you tie with probability 15%, and lose with probability 42%. The probabilities that you win or lose for all possible results are listed in Table 7.3.

Table 7.3 Winning and losing probabilities for different results

Your result	17	18	19	20	21	Blackjack	Bust
P(win)	28%	43%	58%	72%	90%	95%	0
P(lose)	57%	42%	28%	10%	5%	0	100%
P(tie)	15%	15%	14%	18%	5%	5%	0

Note now that the numbers in the table are *conditional* probabilities *given your result*, and in order to compute the probability that you win or lose a single round of blackjack, we need to multiply each conditional probability with the probability of the corresponding case and add them by the law of total probability. Of more interest than "win or lose" is your expected gain. As your actual gain can be either −\$10, \$10, or \$15, we need to figure out the probabilities of these and compute the expected value. The probability that you lose \$10 can be obtained from the two tables above and equals

$$0.15\times0.57+0.15\times0.42+0.14\times0.28+0.18\times0.10+0.05\times0.05+0.28\times1$$
$$\approx 0.49$$

where the term that equals 0 has not been spelled out. You lose with probability 49%, which means that you are actually less likely to lose than not, but because of ties you don't win in all remaining cases. In fact, the probability of a tie is about 10%, so you are still less likely to win than you are to lose. We have already seen that you win \$15 with a probability slightly less than 5%, and a computation similar to the one for losing above reveals that you win \$10 with probability of about 36%. I have used rounded numbers because they are nicer to look at, but to compute your expected gain, I used more decimals to obtain the value −0.53, an expected loss of 53 cents per \$10 wagered, a house edge of 5.3%. This is almost exactly the same as in roulette but remember that this is when you use the dealers's strategy and this is far from the best that you can do. Let us look closer at what kind of information you have and how to use it.

The game starts with the dealer giving you (and all other players) two cards face up and himself one card face down and one face up. The first decision you face is whether to hit or stand after you have been dealt your first two cards. You must make this decision based on the knowledge of your two cards and the dealers face-up card. If you have at most 11, hitting is a no-brainer: as long as it is impossible to go bust, keep hitting. It is when your score is at least 12 that the problems start. Let us say that you have 16. You will then go bust if you hit and the next card is anything between 6 and king and this has probability $8/13$, about 62% (if you get an ace you of course count this as 1, not 11). Should you take this risk? It depends on what the dealer has. Let us say that the dealer shows a 6. This is not a very good card for the dealer because there are a lot of ways he can go bust. By considering all of these, it can be computed that he faces a 42% chance of going bust. Thus, if you decide to stand at 16, the probability that you win is 42%. If you instead decide to hit, your chance of winning is less than 38% because you lose if you go bust, but even if you don't, you can still lose in many other ways. Conclusion: If you have 16 and the dealer 6, stand.

By making considerations like those above, extensive calculations and computer simulations have decided on a "basic strategy" that is the best to follow for those of us who are not Rain Man. In short, if the dealer has 2–6, stand as soon as you risk going bust (that is, stand on 12 or higher) and if the dealer shows 7 or higher, hit until you have 17 or more (the exceptions are if you have 12 and the dealer has 2 or 3, in which case, you should hit). Essentially, it is bad for the dealer to have 6 or lower and good to have 7 or higher and you should adjust your willingness to risk going bust accordingly. One obvious effect of this strategy is that you will go bust less frequently than with the dealer's strategy. It can be calculated that the probability to go bust drops from 28% to 17%. For the uninitiated, it may seem weird to stand already at 12, but it's the right thing to do if the dealer has a bad card. Note also that 12 is no worse than 13, 14, 15, or 16 because the dealer must keep hitting until he gets at least 17, so if he can't beat 16, he can't beat 12 either.

The basic strategy can on occasion be very stressful. Suppose that the dealer shows 7 and you have just reached 16 through the sequence 2–5–2–3–4. It then certainly takes nerves of steel to calmly ask for yet another card. You are not in a very good position to win, but if you lose, at least you did so following the best possible strategy. Your table neighbor who won after standing at 15 can scoff all he wants, you will beat him in the end. Alas, you will still not beat the house: The basic strategy gives the house an edge of less than 0.5% but still an edge.

The basic strategy also gives rules for when to double down, that is, double your wager after your first two cards. For example, if your first two cards add up to 11, you are in a very good position. You have a 31% chance to get 21; in which case, the best the dealer can do is to tie and your chances to get other good totals such as 19 or 20 are also decent. If, in particular, the dealer has one of his bad cards, for example 6, you have a very good chance to win and it makes sense to double your wager. After doubling, you only get one card so this is one of the few situations in any card game when you really don't want an ace. The case of your 11 versus the dealer's 6 is the strongest position you can have and it can be calculated that you have an expected gain of 36% in this particular case. The basic strategy tells you always to double at 11 unless the dealer has an ace; for other scores whether you should double depends on what the dealer has. The weakest position in which the basic strategy advices you to double is if you have ace and 2 (a so called *soft* 13 that can be counted as either 3 or 13) and the dealer has 5. In that case your expected gain is only 0.003% or 3 cents per $1,000 wagered. Not much but still an expected gain and how often do you get that in a casino?

Finally, the basic strategy offers advice for when to split (your cards that is; it does not tell you when to pack up and leave). If your first two cards constitute a pair, you are allowed to split them up into two separate hands by placing another wager on the table. You then play these two hands one after another. One instance in which you should always split is if you have two eights. First of all, 16 is not a very good hand, and second, starting from a single 8 is pretty good. If you have two sixes, 12 is not a very good hand either, but neither is starting from a single 6, so in this case you should only split your pair if the dealer has one of his bad cards, at most 6. You should also always split aces but never tens. That might make sense, but some splitting rules are not intuitively clear. For example, you should split fours if the dealer has 5 or 6 but not otherwise. Not even the Chevalier de Méré would have been able to come up with that rule based on persistent gambling.

There are some further rules and exceptions, and these also vary from casino to casino, but in essence blackjack is one of the fairest (read: least unfair) casino games, and with the basic strategy, you can feel very clever when you win and merely unlucky when you lose.

MATH FOR LOSERS

I have met probabilists who refuse to place even a minuscule bet at a roulette table with the argument that the expected gain is negative. I find this to be slightly on the ridiculous side. Not that they will not gamble, that is of course anybody's decision to make and I certainly do not gamble often myself (I live only a few minutes' drive from a casino and I have not even been there). I know people who think betting is simply a stupid way to spend your money and that is fine. It is the notion that the negative expected value would be the reason that I find a little bit silly. After all, we all know that, don't we? People buy lottery tickets and place roulette bets for their amusement and for the chance to win, not to make a living. If we always lived by expected values, we should never buy insurance either because that is an expected loss as well. I enjoy gambling occasionally because it gives an opportunity to watch randomness and the law of averages unfold in front of you like no other activity. OK, OK, I admit that winning is more fun than losing. In this section, we will assume that the law of averages does not deter us from entering the casino and once we are there, at least trying to gamble as intelligently as we can. If you still find this morally reprehensible, there are plenty of online options to gamble for fun without risking any money. Let's go.

Imagine the following situation. You desperately need \$200 to catch the night flight to Los Angeles (make up your own story why) but you only have \$100. However, there is a casino next to your hotel and you decide to risk your \$100 for the chance to double your money. You know that the odds are against you, but the \$100 you have are worthless to you if you can't afford that flight ticket. How should you bet your money? You could go in for all-or-nothing and put your \$100 on any even-money bet that gives you a probability of $18/38 \approx 0.47$ to instantly double your money and head off for the airport. If you are the cautious type, you may decide to wager only half your money in order to still be in the game if you lose. If you place repeated wagers of \$50, what is the probability that you manage to reach \$200 before you go broke? With probability $0.47^2 \approx 0.22$, you succeed already after two rounds, with probability $0.53^2 \approx 0.28$ you go broke already after two rounds, and with the remaining probability ≈ 0.5 (slightly less than 0.5 to be exact) you are back where you started after one win and one loss. Aha, you think, a case for the recursive method to solve for the unknown probability. Yes indeed, and doing this shows that the probability to succeed with this more cautious strategy is about 0.45. Risking it all gives you a 47% chance to succeed, risking \$50 at a time only 45%. And if you start risking even smaller amounts, you will do even worse. The buzz words in a case like this are *bold play*: Bet it all and hope for the best. The intuition behind this is that as the game is unfavorable to you, expose yourself to the unfairness as little as possible. The house makes its profit in the long run, so the longer you play, the worse you are off (just like when George Foreman lost his heavyweight boxing title to Muhammad Ali in Zaire in 1974; he kept pounding and pounding on Ali's arms and shoulders, exhausting himself and being handily knocked out in round eight). The worst strategy in an unfavorable game is *timid play*, that is, to place minimum bets. With this strategy, you are not likely to make it to LAX tonight.

The situation above can be described as a random walk just like Tom and Harry's coin tossing game in Chapter 6. You start with $100 and place repeated bets of the same size and play until your fortune hits either $0 or $200 (let us assume that your bets are of a size such that you can hit $200 exactly; it is of no interest to you to exceed your goal). Let us look at this in general. Suppose that you have an initial fortune of $\$a$ and set a goal of reaching a fortune of $\$b$ and then stop gambling. You may of course fail to achieve your goal if you go broke first. You will thus keep gambling until your fortune has hit either $0 or $\$b$. You bet $1 at a time, an assumption that is no restriction since we can always replace "dollar" by "betting unit." For example, if you start from $100 and try to reach $300 by betting $50 at a time, your betting unit is 50, you start with two units, you are four betting units away from your goal and two away from ruin. The probability that you succeed is the same as in the one-dollar game with $a = 2$ and $b = 6$. Also let p be the probability that you win a single round and let $q = 1 - p$, the probability that you lose. In a fair game $p = q = 0.5$ and in an unfair game $p < q$. What is the probability that you succeed in reaching your goal?

The fair game is easier because of the following observation: If you start with a units, your expected fortune after each round is always a units. Of course your *actual* fortune goes up and down but because the game is fair, your fortune is on average always equal to what you started with. And the same is true for that moment when you finally reach your goal or go broke.[3] When this happens, you are either at 0 or b. Let p_b be the probability that you

[3]Actually, this is not obvious but relies on a sophisticated result called the *optional stopping theorem* for martingales (recall footnote 1 on page 183). The decision to gamble until you reach your goal or go broke is said to define a *stopping rule*, and it depends on the stopping rule whether the expected value is still a. For example, with the stopping rule to simply gamble until you go broke, your expected fortune at the time this happens is 0, not a. However, it

succeed to reach b before 0; the probability that you fail and go broke first is is then $1 - p_b$. As the expected value is a we get the equation

$$a = b \times p_b + 0 \times (1 - p_b) = b \times p_b$$

and the solution is simply

$$p_b = \text{probability to reach } b \text{ before ruin in a fair game} = a/b$$

Trying to double your fortune corresponds to $b = 2 \times a$ and the probability that you succeed is $1/2$, which makes sense because the game is fair. This also means that in a fair game, there is no advantage to bold play. In the situation above, where you want to double your \$100, bold play corresponds to $a = 1$ and $b = 2$. If you instead bet \$50 at a time, this is $a = 2$ and $b = 4$, ten at a time is $a = 10$ and $b = 20$, and so on. Regardless of how you play, you have a fifty–fifty chance to succeed. If your goal is instead to quit as soon as you are ahead, this corresponds to $b = a + 1$ and the probability that you succeed is $a/(a + 1)$.

The unfair game is both more interesting and more difficult. We now have $p < q$, and the random walk has a tendency to go down, so your expected fortune is a little smaller after each round of the game. It turns out that the crucial quantity is the ratio $q/p > 1$, and it is possible to show that

$$\text{probability to reach } b \text{ before ruin in an unfair game} = \frac{(q/p)^a - 1}{(q/p)^b - 1}$$

This formula is of course not intuitively clear at all, so let us try it with some numbers from your roulette game above. There, $p = 18/38$ and $q = 20/38$, which gives $q/p = 20/18 = 10/9$. Different possible strategies can now be described by changing a and b according to what I said above about betting units. Let us compare three strategies:

bold play: $a = 1$ and $b = 2$

intermediate play: $a = 10$ and $b = 20$

can be shown that your stopping rule in the roulette game we are dealing with indeed gives expected value a.

timid play: $a = 100$ and $b = 200$

Plug in these values in the formula above to find the following probabilities to succeed:

bold play: 47%

intermediate play: 26%

timid play: 0.003%

The revenge for the timid player comes if you instead want to quit while you're ahead. This corresponds to letting $b = a + 1$ for the three a values 1, 10, and 100 and gives these probabilities to succeed:

bold play: 47%

intermediate play: 85%

timid play: 90%

In bold play, you always bet all or nothing, so there will always be just one round and "doubling" and "getting ahead" become the same. For the other strategies, you have the chance to recover and reach your goal even after a number of losses as long as you have money left to bet. Note that you actually have a very good chance to quit while you are ahead with timid play. This may seem surprising because we know that the house cannot be beaten. The crux is that when you are ahead, you have only gained $1 whereas when you go broke, you have lost $100. The house edge remains unchanged.

It is interesting to use the formulas to compare a fair game with one that is only slightly unfair such as craps. We saw earlier that you have a 49.3% chance of winning a pass line bet, and this is very close to fair. Start with $100, bet $1 at a time, and try to double your fortune before going broke. In a fair game, you have a 50% chance to succeed, but in craps the formula above with $p = 0.493, a = 100$, and $b = 200$ reveals that your chance to succeed is only 6%. And this in a game where the success probability is so close to 50%. If you look back at the graphs of Tom and Harry's fair game, you recall that the paths are very likely to stray away one way or the other, stay away for perhaps a long time, but eventually always return to the starting point. An unfair game produces a path that will only return to the starting point a finite number of times and then stay on the negative side forever, relentlessly pushing lower and lower. Unfair games are so unfair.

The formula for the unfair game was used to compute the probabilities for Chevalier de Méré in the beginning of this chapter. In the first case, he tries to get at least one 6 in four rolls of one die; the game is favorable to him, so the formula applies to the house. When he instead rolls a pair of dice 24 times and tries to get at least one double 6, the game is unfavorable to him. If you try the formula on your own, be sure to use exact values of p and q because the final results can change quite a bit due to round-off errors.

Let us return to the casino and to your roulette game. You are joined by a lady who is also going to Los Angeles, but her ticket is already bought and she is only there to kill some time. The croupier announces that he is about to close soon and asks how many more rounds you wish to play. If your new lady friend decides to place repeated \$1 bets and wants to maximize the probability that she ends up ahead, how many rounds should she choose? This is a different situation from above. Although she has a good chance of being able to quit ahead if she plays timidly, she would not know in which round she gets ahead. As it turns out, she should instead bet 35 times on a single number. Why?

As you know, the payout on a straight bet is \$35, so she will be ahead if she manages to win at least once in 35 rounds. If she wins exactly once, she has lost \$34, won \$35, and kept the \$1 she wagered on the winning spin. She has \$36 dollars, and is ahead by one. If she wins more than once, even better. She will be ahead unless she loses all 35 rounds, and the probability of this misfortune is

$$\text{P(lose 35 rounds)} = (37/38)^{35} \approx 0.39$$

and thus she has a 61% chance to be ahead after 35 rounds. This may be a bit surprising. She is actually more likely than not to be ahead; yet this is an unfavorable game. A flaw somewhere? No, she still faces an expected loss because if she fails to win any rounds she has lost \$35, if she wins once she has won only \$1, and together these make up 77% of all cases. If she wins twice, of which there is a 17% chance, she is up \$37; if she wins three times, she is up \$73, but there is only a 5% chance of this; and so on. The probabilities rapidly decline and the larger amounts that she can win cannot compensate the quite likely \$35 loss. She is more likely to be ahead than behind, but the usual 5% expected loss prevails and no strategy in the world can ever improve that. This is one of many examples that illustrate the difference between the probability to be ahead and the expected gain.

What if she instead plays just one additional round, 36 instead of 35? Then one win is not enough to get ahead; she would only break even. To get ahead, she would need at least two wins. The probability that she has no wins after 36 rounds equals

$$\text{P(no wins after 36 rounds)} = (37/38)^{36} \approx 0.38$$

and the probability to get exactly one win is

$$\text{P(one win after 36 rounds)} = 36 \times 1/38 \times (37/38)^{35} \approx 0.37$$

and the probability to get at least two wins is obtained by adding these two and subtracting from one, giving the result 0.25, so after 36 rounds, she only has a 25% chance to be ahead. On the other hand, with two wins, she will be ahead also after 71 rounds (start with \$71, lose \$69, win $2 \times \$35 = \$70$, and keep the two that gave the wins; a total of \$72, ahead by \$1), so let us compute the probability of this. The probability of no wins after 71 rounds is

$$\text{P(no wins after 71 rounds)} = (37/38)^{71} \approx 0.15$$

and the probability to get exactly one win is

$$\text{P(one win after 71 rounds)} = 71 \times 1/38 \times (37/38)^{70} \approx 0.29$$

so that this time the probability of at least two wins is 0.56. The 56% probability to be ahead after 71 rounds is smaller than the 61% chance she has after 35 rounds. You realize that her best choice must be to keep playing until right before she reaches the "breaking point," at which she needs one additional win to be ahead. The next suggestion would therefore be $3 \times 36 - 1 = 107$ rounds, which would require at least three wins and the probability of this is 54%, still a little less likely, and this pattern of decline continues. This is not so surprising because in order to be ahead, she must win on average once every 35 rounds and we know that she only wins on average once every 38. It gets more and more difficult to achieve her goal the more rounds she decides to play. Thus, if she places straight bets, the best she can do is to play 35 rounds for a 61% chance to be ahead. Figure 7.1 gives her probability of being ahead as a function of the number of rounds. Notice the sawtooth pattern with the consecutive peaks at 35, 71, 107, and so on. The peak of 0.61 at 35 rounds is the highest of them all, and they keep getting lower and lower after that. You cannot see it in the figure, but the peaks approach 0 because in the long run, it is impossible to keep winning every 35 rounds with a success probability of $1/38$.

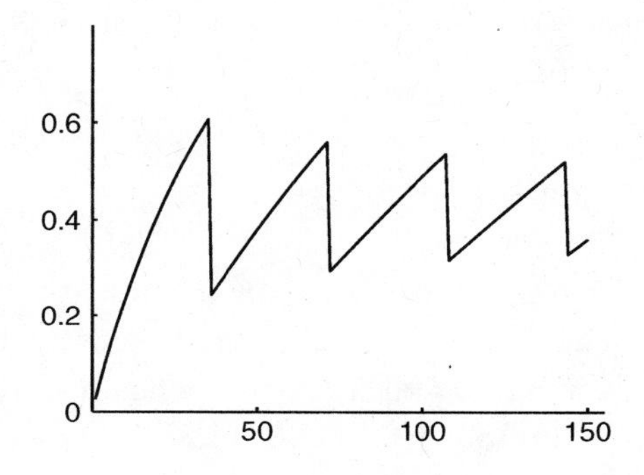

Figure 7.1 The probability to be ahead as a function of the number of rounds of straight roulette bets. The highest probability is 0.61 and occurs after 35 rounds.

What about other bets? If she places one bet on odd, she will be ahead if she wins, which has probability $18/38$ or about 47%, significantly less than 61%. If she places three bets, she needs to win at least two of them, which has probability 46%. With five bets, she needs to win at least three, which has probability 45% (four bets is worse because she still has to win three of them). The best she can do with an even-money bet is to play just once. I suppose it is now easy to accept that any other bet will give a probability between 47% and 61%, but let us check one to make sure. Suppose that she decides to go with street bets, betting on three numbers. The payout is 11:1, so if she plays 11 rounds she will be ahead if she wins at least once (and in line with the results above, this is the best she can do with this bet). The probability to win a single round is $3/38$, and the probability to be ahead after 11 rounds is as usual obtained by first computing the probability that she loses all 11 rounds, then subtracting this from one. We get

$$P(\text{ahead after eleven rounds}) = 1 - (35/38)^{11} \approx 0.595$$

so she has a 59.5% chance, not quite as good as with the straight bets.

Let us finish with a general formula. If she makes a bet on n numbers, the probability that she loses a single round is $1 - n/38$ and the payout is $36/n - 1$. The best she can do is to play $36/n - 1$ rounds, and if she wins at least once, she ends up ahead. The probability of this is

$$\text{P(end up ahead)} = 1 - (1 - n/38)^{36/n-1}$$

and you can check that this is largest for $n = 1$, the straight bet, and declines as n increases.

WIN MONEY AND LOSE FRIENDS

The only chance you have to consistently make money from gambling is to play the role of the house. You need to be able to present a game to make it sound fair or even favorable to your opponent when it in reality favors you. Recall the surprising result from page 54 that there is a better than 50% chance that at least 2 people in a group of only 23 share birthdays. You can use this insight to offer your friend Albert a side-bet when you're both losing money at the roulette table. I am not suggesting that you go around and ask people for their birthdays but instead use the roulette table and suggest to Albert that you both wager $1 and if any number is repeated within the next eight spins of the wheel you win, otherwise he wins. Of course he will take it. There are 38 numbers to choose from, so in only eight spins, surely they are most likely to be all different. But surprise, surprise to Albert, he will lose his money on this bet.

Just like in the birthday problem, let us compute the probability that all numbers are different and subtract this from one. Imitate the calculation in the birthday calculation to get

$$\text{P(some number repeated)} = 1 - \frac{37}{38} \times \frac{36}{38} \times \cdots \times \frac{31}{38} \approx 0.55$$

so you have a 55% chance to win and thus an expected gain of 10%. Now, eight is the smallest number you need to have an edge, but you can probably get away with more at least in the beginning. Suggest ten instead of eight spins. A number repeated in only ten spins with 38 numbers to choose from? Surely Albert will buy it, but the probability that you win is 73%. And if you suggest 19 spins, half of 38, you win with probability 99.6%. What poor Albert doesn't lose to the casino, he will lose to you.

I will now describe some games that can be presented to family and friends as if they were fair but in reality are stacked in your favor. The first can be described as a random version of *rock-paper-scissors.* In case you have forgotten this noble game, you and your opponent simultaneously show either rock (clenched fist), paper (open hand), or scissors ("V sign") with one of your hands. The rules are that rock breaks scissors, scissors cut paper, and paper covers rock.[4] Now, if you knew what your opponent would show, you would of course always win just like Lisa beats Bart on an episode of *The Simpsons*, knowing that "poor predictable Bart, always picks rock" (and Bart's thoughts are "Good ol' rock. Nothing beats that!"). Even if you don't know exactly, you might have an idea of how your opponent plays. Perhaps you are up against a "rock man" who tends to shows rock more often. You counter this by showing paper more often. Or maybe your opponent is a "paper lady," in which case, you show more scissors and so on. If you and your opponent were presented with a choice of strategies (probability distributions on the set of three choices) before each round, you could make sure to win in the long run if you knew your opponent's strategy and chose yours accordingly. We will use this idea to construct a favorable game.

Instead of playing with fists, we will construct three dice, A, B, and C, numbered on their six faces in the following way:

A: 1, 1, 5, 5, 5, 5

B: 3, 3, 3, 4, 4, 4

C: 2, 2, 2, 2, 6, 6

These dice are such that if they are rolled against each other two by two, on average A beats B, B beats C, and C beats A. It is easy to see that A beats B because it does so if it shows 5, which has probability $2/3$. Likewise, B beats C whenever C shows 2, which also has probability $2/3$. Finally, C beats A if either C shows 6, which has probability $1/3$, or if C shows 2 and A shows 1, which has probability $2/3 \times 1/3 = 2/9$. Add these numbers together to find that C beats A with probability $5/9$. We have

$$P(\text{A beats B}) = 2/3 > 1/2$$

[4]That is, those are the most common rules. In the *Seinfeld* episode "The Stand-In," Kramer and Mickey play according to the rules that rock beats anything and their game is not easily decided.

P(B beats C) $= 2/3 > 1/2$

P(C beats A) $= 5/9 > 1/2$

The last probability 5/9 is about 56%, so your advantage is smaller than in the other two cases where the probability is about 67%. The dice game is similar to rock-paper-scissors in the sense that the game is *nontransitive*: In the long run, A beats B, B beats C, but C still beats A.[5] And now it's clear how to siphon off Uncle Sid's money after Christmas dinner: Ask him to play the dice game and be courteous enough always to let him choose his die first. If this makes him suspicious (it will after a while), you can take turns in choosing first. It might take some time before he learns how to use your choice to his advantage, but you will always know how to use his. To make the game less transparent you can construct more complicated dice with more faces.

In the sense I have described above, there is no best die. However, if all three dice are rolled, die A is most likely to win, followed by C, and with B in last place. The reason for this is that A wins whenever it shows 5 and C shows 2 and this has probability $2/3 \times 2/3 = 4/9$ or about 44.4%. Die C wins if it shows 6, which has probability 1/3 or 33.3%, and die B wins if A shows 1 and C shows 2, which has probability about 22.2%. It is hard for die B to beat both the others, but it still beats die C most of the time. In fact, every time A wins, B comes in second.

A variant of the dice game that is sometimes presented is to consider three political candidates, let us call them Al, George, and Ralph. Suppose that one third of voters rank them in the order Al, George, Ralph; one third in the order George, Ralph, Al; and one third in the order Ralph, Al, George. Then Al can truthfully claim that two thirds prefer him to George, and Ralph can claim that two thirds prefer him to Al. However, George is still preferred over Ralph by two thirds of the voters. Political preference is another example of a nontransitive relation, and this is relevant in a multi-candidate election that leads to a runoff between the top two candidates. If Al went to the runoff, he would be afraid to face Ralph.

[5]In mathematics, a *transitive relation* R has the property that if xRy and yRz, then also xRz. A typical example is directed inequality: If $x>y$ and $y>z$, then $x>z$. Undirected inequality is not transitive though because we can have $x \neq y$ and $y \neq z$ but still $x=z$. In rock-papers-scissors, the relation R is "beats," and in our dice game, the relation is "beats with probability greater than 50%."

When Uncle Sid has figured out how the dice game works against him, you can suggest a variant. You will use die A and let him use die C, but instead of rolling just once, you will each roll twice and compute your total score. As C is better than A in one roll, surely it is better in two rolls as well? Of course not. Your possible outcomes are 2, 6, and 10, which have probabilities $1/9$, $4/9$, and $4/9$, respectively, and Uncle Sid can get 4, 8, or 12 with probabilities $4/9$, $4/9$, and $1/9$. You win if (a) you get 6 and he gets 4 or (b) you get 10 and he gets 4 or 8. Adding the probabilities of the two scenarios (a) and (b) gives

$$\text{P(you win)} = 4/9 \times 4/9 + 4/9 \times 8/9 = 48/81 \approx 0.59$$

so although Uncle Sid has a 56% chance to win each single round, his chances are only 41% to win when you add over two rounds. This is similar to the phenomenon of the mutual fund and the fixed interest scheme from page 120.

When Uncle Sid has tired of losing money on dice you can suggest the coin tossing game *Penney-ante* instead. This game was first described by mathematician Walter Penney in a 1969 article in *Journal of Recreational Mathematics* (and regardless of what you have been told, recreational use of mathematics is harmless). Suppose that you toss a coin three times. There are then eight equally likely patterns:

HHH, HHT, HTH, HTT, THH, THT, TTH, TTT

so if you and Uncle Sid were to choose and bet on one pattern each and toss the coin in sequences of three, this game would be fair. Let us change this a little bit. Instead of tossing the coin in sequences of three, toss it repeatedly and wait for the *first* occurrence of either of your patterns. For example, suppose that you have chosen HTH and Uncle Sid has chosen THT. If the sequence of tosses is as follows, you lose after 10 tosses:

HTTHHTT<u>THT</u>

Note that this is different from tossing the coins in sequences of three. In that case, you would not have lost because the first three sequences were HTT, HHT, and TTH, and the fourth sequence starts with T.

Is Penney-ante fair? Surprisingly the answer is that it is not necessarily so and you can use this insight to your advantage. The trick here is that although each triple is equally as likely to show up in any particular sequence of three tosses, some patterns are still more likely than others to show up first in a long sequence of repeated tosses. This seems paradoxical but is easily illustrated with an example. Suppose that Uncle Sid chooses the pattern HHH and you choose THH. The only chance he has to win is then if the very first three tosses give heads and this has probability $1/8$. If this does not happen, you win. Think about it. If there is a T anywhere among the first three tosses of the sequence and you toss until HHH first comes up, it must be preceded by THH and you have won. Thus, if Uncle Sid chooses HHH, you should choose THH for a $7/8$ or 87.5% probability to win. Great odds. But what if he chooses some other pattern?

The beauty of Penny-Ante is that regardless of which of the eight patterns he chooses, you can always choose your pattern so that the probability that you win is at least $2/3$! The basic rule for how to choose your pattern is to let the *first* two letters in his pattern be the *last* two letters in yours. The intuition behind this strategy is that whenever his pattern has a chance to come up after the next toss, chances are that yours is already there. For example, suppose that he chooses HTH. Since his first two letters are HT, you should make these your last two, choosing either THT or HHT. You probably agree with me that THT and HTH must be equally likely to come up first because they look the same, only with heads and tails exchanged, and you should therefore choose HHT. To prove that this gives you an edge over Uncle Sid, we will once again resort to the recursive method and consider a few different cases for the initial tosses.

If the first toss gives tails, you start over because both your patterns start with H. In the case when the first toss gives heads, let us look at the second toss. If this also gives heads, you win because if the coin tossing sequence starts with HH, it is impossible to get Uncle Sid's pattern HTH without your pattern HHT showing up first (try it!). In the remaining case, the sequence starts with HT and two things can happen in the third toss: (1) it gives heads and Uncle Sid's pattern HTH has come up and you lose already after the first three tosses or (2) it gives tails and the game starts over. See Figure 7.2 for a tree diagram illustrating the four different cases. Convince yourself that these are all the cases and that we have not forgotten anything.

What is now your probability to win? Let us denote it by p and obtain an equation, the usual deal with the recursive method. We can ignore the case in which you lose, which leaves us with three cases: T, HH, and HTT. The

Figure 7.2 The four different ways to start a game of Penney-ante when Uncle Sid has the pattern HTH and you have HHT

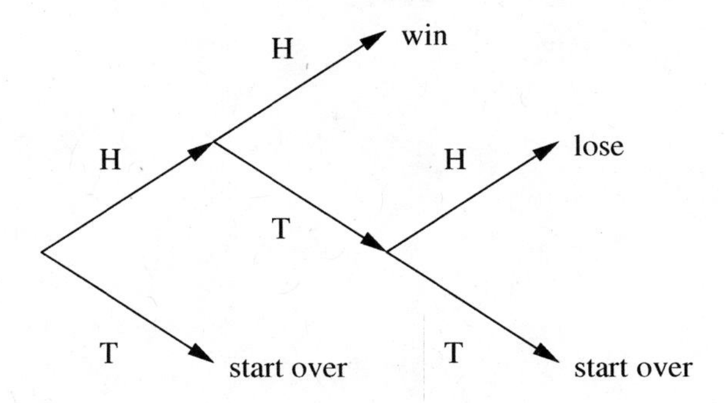

conditional probabilities that you win in these cases are

$$\text{P(win given T)} = p$$

$$\text{P(win given HH)} = 1$$

$$\text{P(win given HTT)} = p$$

and because the cases have probabilities P(T) $= 1/2$, P(HH) $= 1/4$, and P(HTT) $= 1/8$, the law of total probability gives us the equation

$$p = p \times 1/2 + 1 \times 1/4 + p \times 1/8$$

which simplifies to

$$p = (2 + 5 \times p)/8$$

which gives $3 \times p = 2$ and hence $p = 2/3$. With probability $2/3$ or about 67%, HHT shows up before HTH and you beat Uncle Sid.

Similar calculations show that whatever pattern Uncle Sid chooses, you can always do better. The complete strategy for you is: Let his first two be your last two and never choose a palindrome (a pattern that reads the same forward and backward). For example, if he chooses HTH, you have a choice between THT and HHT, and because THT is a palindrome, discard it and go with HHT. Table 7.4 gives your best choice of pattern and the associated winning probabilities.

Table 7.4 Uncle Sid's choice and your best response in Penney-ante

Uncle Sid	You	P(You win)
HHH	THH	87.5%
HHT	THH	75%
HTH	HHT	66.7%
HTT	HHT	66.7%
THH	TTH	66.7%
THT	TTH	66.7%
TTH	HTT	75%
TTT	HTT	87.5%

Just like the dice game above, Penney-ante is nontransitive. There is no *best* pattern but always a *better* pattern and Uncle Sid has to pay up.

While Uncle Sid is trying to explain the sudden deficit in the household budget to Aunt Jane, let us look a bit closer at the emergence of patterns in the sequence. A question related to which pattern beats which is how long the expected wait is for a particular pattern. Let us do the simplest case, which is the pattern HHH and again use the recursive method, this time for the expected value. Let us denote the expected number of tosses until we first get HHH by μ and consider some different cases for the first few tosses. If the first toss gives tails, one toss has been spent and we start over expecting an additional μ tosses. In this case, the expected total number of tosses is therefore $1 + \mu$. If the first toss gives heads and the second gives tails, a similar argument gives that the expected total number of tosses is $2 + \mu$. If the two first tosses give heads and the third gives tails, the expected total number of tosses is $3 + \mu$, and if the first three tosses give heads, the expected total number of tosses is simply 3 because we already have the pattern HHH. We have described four different cases: T, HT, HHT, and HHH and after convincing ourselves that there are no other possibilities, we compute their probabilities $1/2$, $1/4$, $1/8$, and $1/8$, respectively, and get the following equation for μ:

$$\mu = (1 + \mu) \times 1/2 + (2 + \mu) \times 1/4 + (3 + \mu) \times 1/8 + 3 \times 1/8$$

which after multiplying both sides by 8 simplifies to

$$8 \times \mu = 14 + 7 \times \mu$$

which gives $\mu = 14$. The expected number of tosses until the pattern HHH first shows up is 14. By symmetry this is also the expected number of tosses until TTT shows up. For the other patterns, similar but more complicated calculations reveal that the expected number of tosses until HTH show up is 10, and by symmetry, the same is true for the pattern THT. All remaining patterns each has an expected waiting time of eight. Note that your choice of pattern in the optimal strategy against Uncle Sid is always among these remaining four. The worst he can do is to choose HHH or TTT; he is then easily beaten. The second worst is to choose HTH or THT, and his best choice is any of the remaining four patterns, but even then, you can choose your pattern so that you beat him two times out of three.

It may still seem paradoxical that each pattern is equally as likely to show up in any particular sequence of three tosses; yet, we have to wait on average 14 tosses for HHH but only 8 tosses for THH. Does this not indicate that THH tends to show up more often in a long sequence of tosses? No, and the revenge for HHH is that once it has come up, there is a 50–50 chance that it comes up again after the next toss if this also gives heads; in the sequence HHHH, there are two occurrences of HHH, so this pattern has the ability to repeat itself immediately. No such repetitive powers in the pattern THH; once this has appeared, we must wait at least three more tosses before it can show up again. And this explains the apparent paradox because the pattern HHH compensates its tardiness with its ability to repeat itself. In a long sequence of coin tosses, the pattern HHH thus tends to show up in bursts, whereas THH is more evenly spread out. In the long run, they show up equally as often but not in the same way.

Now, HHH and THH were just examples chosen to represent each category above. Thus, the "slow" patterns HHH and TTT can repeat themselves immediately and the "fast" patterns HHT, HTT, THH, and TTH must start over from scratch in order to reappear. The remaining patterns HTH and THT are intermediate. They cannot repeat themselves immediately, nor do they have to start over completely but can repeat themselves after two tosses. For example, there are two occurrences of HTH in the sequences HTHTH. The key concept turns out to be how much *overlap* a sequence has with itself. The pattern HTH has an overlap of length one because the last H can be the first H in another occurrence. The same is true for HHH, but this pattern also has an overlap of length two because the last two Hs can be the first two in a new occurrence. The remaining sequences have no overlap at all.

There is nothing magical about the number three (in this context), and Penney-ante can be played with patterns of any length n and the basic strategy remains similar to the one I described above. As long as $n \geq 3$, you can always beat Uncle Sid; in the case $n = 2$, you cannot always choose a better pattern than Uncle Sid. If he chooses HT or TH, the best you can do is to counter with TH or HT, respectively, for a 50–50 chance to win. If he chooses HH or TT, you beat him with TH or HT, respectively. In the worst two cases, you are equally as likely to win as you are to lose, and in each of the other two, you win with probability $3/4$ (try to compute this on your own).

For a general pattern of length n, the same principle as in the case of length three applies: The more overlap the pattern has with itself, the longer we have to wait for it to appear. Recall my "unremarkable" pattern

HTTHHTHTTHTTTHHTHTTT

from page 84. This is certainly considered less remarkable than if you would get 20 heads in a row:

HHHHHHHHHHHHHHHHHHHH

even though both patterns have probability $(1/2)^{20}$ to show up in a particular sequence of 20 tosses. However, my sequence is much more likely to appear before the 20 heads in a row in a sequence of repeated tosses, so in this sense it is actually less remarkable. The pattern with 20 heads compensates for this by plenty of overlap and can repeat itself immediately and then again and again, whereas my pattern has no overlap at all and we must start over and wait at least another 20 tosses before it can show up again. Many patterns that we would consider special or remarkable would be those that have some particular repetitive feature such as

HTHTHTHTHTHTHTHTHTHT

which has overlaps of lengths 2, 4, etc., or

H H H H H T T T T T H H H H H T T T T T

which has an overlap of length 10, and these therefore tend to be a little slower to appear. In contrast, the unremarkable patterns with no overlap at all are more likely to show up earlier. Some patterns without overlap may still be considered remarkable such as H H H H H H H H H H T T T T T T T T T T, ten heads followed by ten tails. Although it cannot repeat itself in fewer than 20 tosses, it can repeat a similar pattern with H and T exchanged in 10 tosses, and it is reasonable to extend our "definition" of remarkable patterns to also include such scenarios.

For any pattern of any length n, there are explicit formulas for how to compute the expected waiting time until it shows up. A very good overview of this and plenty of other good stuff for the mathematically inclined can be found in the 1994 book, *Problems and Snapshots from the World of Probability*, by Swedish trio Gunnar Blom, Lars Holst, and Dennis Sandell. From this we learn that you have to wait on average a little over one million tosses for my unremarkable pattern of length 20 (it is after all a million-to-one shot) and twice as long for a sequence of 20 heads. The formulas are surprisingly simple, but rather than stating them generally, I will illustrate how they work in a few examples. Say that a sequence has overlap of length k if the k last digits equal the k first. For example, the sequence H H T H H H has one overlap of length 1 and one of length 2. It also has one of length 6, which is trivial and true for any sequence of six tosses. Thus, this sequence has overlaps of lengths $1, 2$, and 6, and the expected time before it shows up is then computed as

$$2^1 + 2^2 + 2^6 = 2 + 4 + 64 = 70$$

The sequence H H H H H H has overlaps of all lengths 1–6, and the expected wait for this sequence is

$$2^1 + 2^2 + 2^3 + 2^4 + 2^5 + 2^6 = 126$$

and a sequence with no overlap at all other than the trivial of length 6, for example H H T H T T, will show up after an average of $2^6 = 64$ tosses. Check that this method of computing gives the expected waiting times we obtained above for the Penney-ante patterns of length three.

The formulas can be extended in various ways, for example, to compute waiting times for patterns in sequences with more than two symbols. For example, if a die is rolled repeatedly, you already know that you must wait on average six rolls to get a 6. If we instead wait for the first occurrence of two sixes in a row (the pattern 66), what is the expected waiting time? To compute this, follow the method above but with 6 instead of 2 (we have 6 symbols rather than 2). As 66 has one overlap of length 1 and one of length 2, we get an expected wait of $6^1 + 6^2 = 42$. Note that the probability to get two sixes in any two consecutive rolls is $1/36$, so if we started over between pairs of rolls, we would have to wait on average $2 \times 36 = 72$ rolls. The 42 rolls are expected if we wait for the first occurrence of the pattern 66 in repeated rolls. The pairwise way of counting disregards the double six in the sequence 413667 because it would group this as 41, 36, and 67. For biblical numerologists, finally, the expected wait for the first occurrence of 666 in repeated die rolls is $6^1 + 6^2 + 6^3 = 258$ rolls.

FINAL WORD

If you ask a probabilist about your gambling prospects, you will get a decidedly negative assessment. Games are simply stacked against you, and although you may win occasionally, the law of averages makes sure that you should not choose the roulette table as your career path. Still, games and gambling are extremely interesting to probabilists because they provide a wide variety of probability problems, some of which are very deep with applications far beyond the walls of the casinos.

8

Guessing Probabilities: Enter the Statisticians

LIES, DAMNED LIES, AND BEAUTIFUL LIES?

I'd better admit it right away: This chapter is about statistics. Wait, don't put the book away just yet! I know from years of teaching that most people find statistics about as exciting as watching wet paint dry and as pleasurable as root canal work. When I wrote the proposal for this book, I mentioned how I therefore deliberately wanted to avoid the word statistics in the book title. One of the reviewers pointed out that although it is true that many people perceive statistics as boring, many also perceive probability as scary. Well, people buy scary books, don't they? But if you ever found probability to be a scary subject, I hope that I have helped you overcome some of your fears by now.

My next mission is to show you that statistics is not always that boring, which may be a more challenging task. You may have suffered through one of those college courses where your head started spinning from all the significance levels, type one and two errors, p-values, and questions about whether two samples were heteroscedastic. Or you may think that statistics is all about some guy with thick glasses staring at page after page of census data, trying to decide whether the population of Luckenbach, Texas has changed in the last decade. And sure, statistics can be about all those things (except that census data will not provide much useful information about Luckenbach as those of you who have been there understand). But statistics can also be really interesting, sometimes even exciting. Let us hear what Sir Francis has to say:

> Some people hate the very name statistics, but I find them full of beauty and interest.
>
> Francis Galton, *Natural Inheritance*, 1889

See there, "statistics" in the same sentence as "beauty and interest," I bet you don't see that every day. And if you think statisticians are cold-hearted number crunchers, you should know that Florence Nightingale (1820–1910), The Lady with the Lamp, the personification of compassion and goodness, was actually a pioneer in the use of statistical methods in the health sciences. Her statistical analysis lead to great improvement in mortality rates at civil and military hospitals, and she became a Fellow of the Royal Statistical Society in England as well as an honorary member of the American Statistical Association. How about that?

You have probably heard the famous saying (attributed by Mark Twain to Benjamin Disraeli) that there are three kinds of lies: lies, damned lies, and statistics. To add insult to injury, Darrel Huff wrote a famous book in 1954 called *How to Lie with Statistics* that described how statistics can be used in deceptive ways. As a counterweight to all of this, let Sir Francis continue his thoughts about statistics:

> Whenever they are not brutalized, but delicately handled by the higher methods, and are warily interpreted, their power of dealing with complicated phenomena is extraordinary.

Collecting, tabulating, and describing data are very important activities, but the "higher methods" that Sir Francis talks about enter when the theory of probability is used to analyze data and draw meaningful conclusions, so-called *statistical inference*. In a way, this is quantified common sense. For example, suppose that a pharmaceutical company tests a new drug by giving it to one group of patients and an old drug to another group. If 83% of the patients given the new drug and 67% of those given the old drug get better, can the company conclude that the new drug is more effective? The evidence for such a claim would be that 83 is quite a bit more than 67. If the numbers had instead been, for example, 69 and 67, nobody would claim this to be a substantial difference. Still, the company wants to draw general conclusions about how effective the new drug is, not just in these particular patients. If the same drugs are given to two new groups of patients, maybe the results will be different. Results from probability can then be used to address such issues, and it can be decided when an observed difference is large enough to be meaningful. You also realize that it matters how many patients are in the two groups. If the 83% is five out of six and the 67% is two out of three, you would not be able to draw any conclusions. The law of averages tells you that the results are more reliable the larger the groups you have, but for a more detailed analysis, more sophisticated results must be incorporated and the pharmaceutical industry is indeed a major employer of statisticians. I have lost track of how many of my former graduate student colleagues now work for British–Swedish pharmaceutical giant AstraZeneca.

We have already touched on the subject of statistics in earlier chapters, for example, when we talked about "statistical probabilities." If a meteorologist has observed that a particular set of weather conditions currently in place leads to rain about 25% of the time, he can express this as saying that the chance of rain is 25%. What he has done is *estimated* the probability of rain based on the available data. As he gathers more data, the estimate may change, and as you know from the law of averages, the more data he gathers, the more reliable are his estimates. Much of statistical theory and methodology relies on the law of averages, although this is not always pointed out. For an early example of this insight, in his 2002 book *What Are the Chances?*, Bart Holland tells us about a Mr. J. Koelbel who in 1584 suggested a way to establish the numerical value of the length measurement "foot." He suggested that 16 men be taken when they leave church, "as they happen to come out," and have them line up their left feet in a row and measure the total length of their feet. This number divided by 16 would then be the "right and lawful foot." Koelbel realized the importance of averaging to smooth out individual variation and arrive at a generally acceptable definition of a foot.

An interesting example of the usefulness of statistical methods comes from World War II. In 1943, the Economic Warfare Division of the American Embassy in London started analyzing serial numbers from captured German equipment, for example, bombs, rockets, and tanks. The statisticians who worked with these data came up with clever ways to use them to estimate German strength. Let us consider tanks as our example. If the Germans had a total number of N tanks, numbered 1 through N, the problem was how to estimate N based on observed serial numbers. For simplicity, suppose that the Allied had captured three tanks numbered 89, 123, and 150. How should they estimate N? There are many ways to approach this and no universally correct answer, but it is obvious that the estimate of N must be a bit more than 150, the largest observation. It can be calculated that the expected value of the largest of three observations randomly sampled from 1, 2, ..., N is about

Table 8.1 Estimated and actual monthly production numbers for German tanks during World War II

Date	Statistics estimate	Intelligence estimate	Actual number
June 1940	169	1,000	122
June 1941	244	1,550	271
August 1942	327	1,550	342

$0.75 \times N$ (on average the observations are evenly spread out) and as 150 equals 0.75×200, this would lead to the estimate $N = 200$. Note how this estimate is based on probability calculations combined with data; the "higher methods" in full swing. There are ways to further refine the estimate, but let us not go into these technicalities but instead finish the story.

Once the war was over, the statisticians got their answer key, which is an unusual situation in statistics where the true answers are seldom known. It turned out that the statisticians (undoubtedly men with thick glasses) had done really well and outclassed the estimates obtained by British and American intelligence. You can read more about the use of statistics during World War II in the (nontechnical) article "An Empirical Approach to Economic Intelligence in World War II," which was published in the *Journal of the American Statistical Association* in 1947, authored by Richard Ruggles and Henry Brodie who were there when it happened. Table 8.1 gives the data of production numbers from three different months. The estimates by statisticians and intelligence officers are compared with the official figures from the Speer Ministry. Note how the intelligence estimates are gravely exaggerated.

In sections to come, we will look at some common uses of statistics. The review will not be as systematic as what we have done for probabilities; instead, I will pick and choose among what I think are the most interesting problems and applications. Throughout, I want you to keep Sir Francis' wise words in mind: delicate handling and no brutality. OK? Then, let's go.

4 OUT OF 10 LIKE THE PRESIDENT 19 TIMES OUT OF 20

Hardly a day goes by without another opinion poll reported in the media. This week (the second week of May 2006), I learned that seven in ten Americans have changed their driving habits due to high gas prices, that 54% of Canadians are opposed to sending troops to Afghanistan, and that 51% of Americans have an unfavorable view of Tom Cruise. We are presented presidential approval ratings more or less continuously, and before an election, there is a veritable explosion of polls from various polling companies and institutes. Many polls never reach the general public. Political parties and candidates do their own secret polling to investigate what issues are important to people and to develop strategies. The public political polls provide a good way for the polling companies to be seen and hopefully gain a good reputation, but they make much of their money elsewhere, polling for companies. When a company wants to introduce a new product, they may hire a polling company to ask people whether they would buy it. I was once called and asked if I (a) used a particular brand of shampoo, (b) drank orange juice daily, and (c) had a dog. After having dismissed the idea that some company tried to introduce an orange-scented dog shampoo, I concluded that the polling company had simply put together three different polls into one in order to save time and money.

When a presidential approval rating of 40% is reported, what does this mean? First of all, it obviously does not mean we know for sure that exactly 40% of the entire population approve of the president's performance. The only way to know this would be to ask everybody, and the idea behind an opinion poll is that we can't do this, so instead we ask, for example, 1,000 people for their opinions. If 400 of those express approval of the president's performance, it is reported that he has a 40% approval rating. This number is used as an estimate of the true but unknown proportion of the entire population that approve of the president. The quality of the estimate is measured by the *margin of error* (also called the *sampling error*), which is a number that is subtracted from and added to the 40% to give an entire interval rather than a mere number. For example, if the margin of error is reported to be 3%, the approval rating interval is 40 ± 3 or [37,43]. OK, what does this mean then? Are we now certain that the true approval rating in the entire population is between 37% and 43%?

Not quite. When randomness is involved, we can never be certain of anything. We have to be satisfied with a high degree of certainty, and this can be achieved. For example, let us suppose that the true proportion of supporters is 50%. How likely is it then to get an approval rating of 40% or less in a sample of 1,000 individuals? This is very similar to tossing a coin 1,000 times and asking for the probability to get less than 400 heads. A computation yields that the probability of this is about one in ten billion so if we get 40% in our sample, it seems extremely unlikely that the true proportion would be as high as 50%. In an opinion poll, the situation is a bit backward because we do not know the true proportion, only what we actually observe. The question then becomes how likely we are to catch the true proportion with our interval composed of the number we observe and its margin of error. Let us look at this from a slightly more theoretical point of view.

Call the true proportion p, now a number between 0 and 1, and our observed proportion $\widehat{p}$ (it is customary in statistical theory to use "hats" to denote that something is estimated from data). If the margin of error is ϵ, the question is how likely it is that the interval $[\widehat{p} - \epsilon, \widehat{p} + \epsilon]$ contains p, and this can be computed by using, for example, the central limit theorem. We then choose an ϵ that makes the probability as high as we desire but note that the higher this probability is, the larger must ϵ be. There is a tradeoff here because we also want the margin of error ϵ to be small. The solution is to choose the probability of catching p in advance, and it has become standard to use 95%. This number is called the *confidence level* because it expresses how confident we are that we have managed to catch p in our interval (which is called a *confidence interval*). The formula for our interval turns out to be as follows:

$$p = \widehat{p} \pm 1.96 \times \sqrt{\frac{\widehat{p} \times (1 - \widehat{p})}{n}} = \widehat{p} \pm \text{margin of error}$$

where n is the number of individuals in our sample. The formula means that the interval given by $\widehat{p}$ plus/minus the margin of error has probability 95% to catch the true unknown proportion p. In other words, out of every 20 intervals constructed in this way, on average, 19 catch p and 1 misses. Note that the margin of error decreases as the sample size increases, which makes sense because an opinion poll should be more believable the more people that have been asked. The formula tells us exactly what is the influence of the sample size. The number 1.96 is associated with the probability 95%.[1] If we want

[1]This association is based on calculations involving the normal distribution. In a normal distribution, the probability is 95% that an observation falls within 1.96 standard deviations of

to have greater confidence we must increase this number; for 99%, we must replace 1.96 by 2.58.

Let us use our imaginary but realistic poll of 1,000 people, 400 of whom approve of the president. This gives us $\widehat{p} = 0.4$ and $n = 1{,}000$, and if we put in these numbers above, we get the interval

$$p = 0.4 \pm 1.96 \times \sqrt{\frac{0.4 \times 0.6}{1{,}000}} = 0.4 \pm 0.03$$

or as percentages, 40±3%. Computing this interval is a good example of the "quantified common sense" I mentioned earlier. We want to know what proportion of the population approves of the president's performance but obviously we can't ask everybody. If we ask 1,000 randomly chosen people, we should at least get an idea of what is that proportion. The information in the sample is not perfect, but it is also far from useless. The margin of error together with the confidence level quantifies this idea and tells us that with 95% probability, we will be able to get the true population proportion within the margin of error of the proportion in our sample.

Many opinion polls have a margin of error that is about 3 percentage points, and this is no coincidence. The 95% confidence level is as I said standard, which gives the factor 1.96, and if the observed proportion $\widehat{p}$ is not too far from 0.5, between 0.3 and 0.7 or so, the square root of $\widehat{p} \times (1 - \widehat{p})$ is about 0.5. As 1.96 is approximately equal to 2, the product of the two is approximately equal to 1, which means that the margin of error is roughly $1/\sqrt{n}$ or, as a percentage $100/\sqrt{n}$, which provides a good rule of thumb to remember:

$$\text{margin of error} \approx \frac{100}{\sqrt{\text{sample size}}} \%$$

Note now that if the sample size is around 1,000, which is also typical, this becomes about 3%. The rule of thumb is useful even for proportions that are not close to 50% because it gives the maximum margin of error; the further the proportion is from 50%, the smaller the margin of error. In polls for American presidential elections, the two major party candidates are both close enough to 50% so that the margin of error is about 3% for samples of size around 1,000. Also note that increasing the sample size by a factor reduces the margin of error by only the square root of the same factor. Thus, if n is increased from

the mean, but as I have not revealed how the confidence interval is created and how it relates to the normal distribution, I will not go into any further details.

1,000 to 10,000, the margin of error decreases from 3% to 1%. As polls cost time and money, such an increase of sample size may not be worthwhile and you nowadays see very few polls with more than a few thousand individuals.

We have seen that if you ask 1,000 individuals in the United States for their opinion, you get a margin of error of 3%. What about other nations? China has about four times the population of the United States, how many people would you need to ask there to get the same margin of error? Four thousand? But look at the calculations and formulas above. Where does the population size enter? Nowhere! A poll of 1,000 people has the same margin of error and is thus just as valid whether it is taken in China, the United States, Canada, or Mexico! This may seem a bit surprising at first, and as a matter of fact, it is not entirely true. Some approximations are involved in the computation of the margin of error, but they are valid as long as the sample size is small relative to the population size. Thus, 1,000 Chinese are just as representative of their country as are 1,000 Americans of theirs. However, if we were to ask 1,000 people in Vatican City, we would have asked everybody and the margin of error would be zero (I don't know if papal polls are being published, but I assume that anything below 100% approval rating would be considered disastrous).

In the Fall of 2005, it was reported in the media that President Bush's approval rating was below 40% for the first time ever. In a way this is a meaningless statement. Although it may be true that one poll had his approval rating at 41% and the next at 39%, these numbers both have margins of error of 3%, and if you construct the corresponding confidence intervals, you will notice that they overlap. The numbers 41% and 39% without their margins of error do not tell the whole story; not until we view them as the intervals [38,44] and [36,42] do they tell us anything relevant. Thus, we cannot even exclude that support in fact *increased* between the two polls. When a difference is so large that it is entirely outside the margin of error, it is said to be *statistically significant*, and unless a difference is in this way significant, there are no conclusions to be drawn although reporters and pundits love to overanalyze poll numbers and talk about trends and tendencies even when there aren't any. And conversely, politicians who receive low poll numbers often comment that they get a completely different "feeling" when they are out "talking to people." It is important neither to over- nor underemphasize opinion polls. They are what they are, no more and no less.

If statistical significance has been established, this is of course still not the same as truth. If we have a 95% confidence level, there is still a 5% probability that an observed statistically significant difference is not real but due to chance alone. However, if we start thinking too much about this, opinion polls may seem to be completely meaningless and they are not. Let us be content with "statistical truth."

The margin of error is usually reported in opinion polls, but I remember one notable exception. In 1995, Sweden joined the European Union, a move that was preceded by years of debate and a referendum. Shortly before the referendum, one newspaper had the headline "Majority of Swedes favor joining the Union!" This was based on an opinion poll where 50.5% had expressed support for joining, and in order to get a margin of error less than the 0.5% needed to be able to draw the paper's conclusion, we would have a confidence level of only 25%! In other words, only one out of four such polls would succeed in catching the correct proportion within its margin of error. This is obviously useless. I asked the paper's political editor why they didn't publish the margin of error and got the reply, "Oops, we forgot!" Needless to say, this particular paper supported Swedish membership in the European Union.

The confidence level, the measure of how much we can rely on the margin of error, is seldom reported. One exception seems to be Canadian polls where you can see statements such as "The margin of error is 3.1 percentage points, 19 times out of 20." Although well meant, the statement is probably more confusing than clarifying if you don't already know what it means. Nineteen out of twenty is another way of saying 95%, and as far as I know, all polling companies use this number. There seems to be a consensus that 95% certainty is just about right; any lower and the results would not be as reliable, any higher and the margin of error becomes too large. And when it comes to opinion polls before presidential elections, the polling companies compete fiercely with each other about who can "get it right," and it is reasonable that they all use the same confidence level.

In any poll, it is important that the sample is representative of the population of interest, and there are many ways to violate this requirement. When I was in graduate school, my fellow graduate students and I did statistical consulting as part of our education. One of my clients was a surgeon who specialized in a certain type of reconstructive surgery. He wanted to have his results evaluated and had asked for opinions from several other doctors. He had shown them pictures of his patients, and the doctors were asked to grade the quality of the surgery according to various criteria. When I looked at his data, I noticed that most doctors were in agreement but that there was one whose grades were consistently different from the rest. When I asked my client about it, he laughed and replied, "Oh, that's Dr. Karlsson, he is a general practitioner, he doesn't understand any of this!" It turned out that my client had just added Dr. Karlsson to increase his sample. He probably remembered from some statistics course that larger samples were better, but if he really wanted to poll experts for their opinions, adding poor clueless Dr. K would of course not do him any good.

For another example of a nonrepresentative sample, suppose that you take an evening walk in your neighborhood and observe that 14 out of 20 people you meet are walking a dog. Would you conclude that about 70% of your neighbors are dog owners? Probably not. This is an example of *selection bias*; the individuals you meet on the street are more likely to be dog owners than the neighborhood population as a whole, and they are thus not a representative sample. Selection bias is a serious error that can lead to false conclusions, and in the next section, we will look at two famous examples from American presidential elections.

POLLS GONE WILD

Before the presidential election in 1936, the magazine *Literary Digest* published an opinion poll that predicted an easy win for Republican candidate Alf Landon over President Franklin D. Roosevelt: 57% for Landon and 43% for Roosevelt. The *Digest* had a solid reputation after having correctly predicted the winner in each presidential election since 1916, and the poll was based on responses from 2.3 million people. Yes, you read correctly, two million three hundred thousand people! The election result? Roosevelt got 62% and Landon 38% (of those who voted for either of the two), which is one of the largest margins of victory in any presidential election. The *Digest* poll has gone down in history as the worst opinion poll, ever and the magazine went out of business shortly thereafter.

How on earth could this happen? With 2.3 million people, the rule of thumb from above gives a margin of error of only 0.07%, so the predicted numbers ought to be almost certain. Did something happen that made people suddenly change their minds? No, the error is in the methods of the *Digest*. The reliability of the estimate as measured by the margin of error is only valid if we have a *random sample*, meaning that everybody is equally likely to be selected. In theory, if there had been a list of all voters and 2.3 million people had been chosen from this list, the prediction would have been very accurate. But this is not how it was done. The *Digest* made two errors that resulted in heavily biased results.

Their first error was selection bias. When they selected the people to be included in the poll, they used various available address lists such as their own subscription lists, telephone directories, automobile registration lists, and club membership lists. Now, this was during the Great Depression and you would not show up on any such list unless you had disposable income. A young man who had just enrolled in the Civilian Conservation Corps would most likely not spend his daily dollar on a subscription to the *Literary Digest*, nor would a recently laidoff steel worker decide to join the local country club. Cars and telephones were also far less common than today; for example, only 25% of households had a telephone. The selection of individuals favored the rich, and in 1936, they were less likely to support Roosevelt's New Deal than Landon's more restrictive financial policies. This might have been the first election year when there was this kind of divide in the electorate that mattered to the *Digest*'s polls. After all, they had managed to get it right before.

As if the selection bias was not bad enough, the second error was *nonresponse bias*. The *Digest* mailed postcards to 10 million people and based their poll on the 2.3 million cards that were returned. One can imagine that even if the recently laidoff steel worker was to receive a postcard from the *Digest*, he would be far more concerned with feeding his family than filling out the card and mailing it back to the magazine. The bias that was introduced by selection was further enforced by nonresponse, and whereas 2.3 million may seem an impressive number, a response rate of 23% is not. It may be a bit more speculative that nonresponse bias would favor Landon, but a special poll done in Chicago showed that over half of those who responded favored Landon but Roosevelt still got two thirds of the vote in the city. The Chicago poll had a 20% response rate and did not suffer from selection bias because the individuals were chosen from lists of registered voters.

In the 1936 election, the *Digest* faced competition from some new kids on the block. Archibald Crossley, George Gallup, and Elmo Roper were three bright young fellows who had realized that samples must be random in order for results to be reliable. Each of them predicted a win for Roosevelt, and Gallup also managed to correctly predict the *Digest*'s erroneous numbers, a feat that established George Gallup more than anybody else as Mr. Opinion Poll. He also gained this reputation in Europe where he later correctly managed to predict Winston Churchill's defeat in the U.K. elections in 1945 when almost everybody else predicted a Churchill victory.

In the 1936 U.S. presidential election, Gallup predicted 56% for Roosevelt (still a bit off the actual 62%) based on a random sample of 50,000 people.

Moreover, based on a random sample of 3,000 people from the lists used by the *Digest*, Gallup predicted that the magazine would predict 44% for Roosevelt. Gallup realized that a sample of 3,000 would give a good idea of how the 10 million on those lists would vote and he realized that the huge sample size of 2.3 million would do no good because the selection procedure was skewed from the beginning. See Table 8.2 for predicted and actual numbers in the 1936 election.

Table 8.2 Polls and election numbers for Roosevelt in the 1936 election

Source	Roosevelt's %
Election result	62
Gallup	56
Literary Digest	43
Gallup's prediction of *Digest*	44

The unemployed editors of the *Digest* would however get a small revenge on the pesky newcomers 12 years later. The 1948 American presidential election is the time of the second famous erroneous poll, and this time it was Gallup & Co. who got it wrong. You have probably seen the famous photo of Harry Truman holding a copy of the *Chicago Daily Tribune* with the headline "Dewey Defeats Truman" just after Truman had won the election. Crossley, Gallup, and Roper had predicted a victory for Republican candidate Thomas Dewey by a handsome 5–7% margin, and in reality, the results turned out the opposite. What went wrong this time?

The reason for the failure of Gallup & Co. was that the three pollsters had still not managed to get rid of all bias in their sampling. They realized correctly that their polls would be more accurate if they made sure that their samples reflected the population composition. Thus, they would try to get half men and half women and various other population traits such as race, age, and income in their correct proportions in the samples. The polls were conducted by interviewers who visited the selected individuals and asked for their opinions. However, once an interviewer was informed that he had to interview, for example, five white men over the age of 40 in suburban Chicago, he was free to choose whomever he wanted, and there enters the potential bias. For whatever reasons you can imagine (nicer neighborhoods, shinier cars in the driveways to draw attention, housewives who are at home to open the door, etc.), the interviewers tended to interview disproportionally many Dewey supporters. And this was no coincidence because the pollsters consistently overestimated the Republican vote in the elections 1936–1948. Republicans were simply slightly easier to interview and that biased the results somewhat, but only in 1948 was the difference between the parties small enough that the bias made the pollsters actually predict a Republican victory. See Table 8.3 for some numbers regarding the 1948 election results and predictions.

Table 8.3 The 1948 election: Predictions by the three pollsters Crossley, Gallup, and Roper and the actual election result

Candidate	Crossley	Gallup	Roper	Election result
Truman	45	44	38	50
Dewey	50	50	53	45
Others	5	6	9	5

To avoid selection bias as much as possible, the individuals included in the sample must be identified when the poll starts. If interviews are done over the phone, interviewers must talk to the person that is selected, not whoever happens to pick up the phone. If interviews are done in person and nobody is at home at the moment, the interviewer should not ask the neighbor or the mailman for their opinions instead. A modern type of selection bias stems from the fact that polling is often done from phone directories, but more and more people, especially the young, have only cell phones and no land line and they are thus excluded from the samples. It is unclear if and how this affects the outcome of political opinion polls, but it would probably have great impact if the question was about support for a ban on cell phones in public places.

It is harder to avoid nonresponse bias, but the polling companies usually try to contact people several times before they give up. If nonresponse occurs randomly and is not too large, it is not too problematic but if it is believed to skew the results, it could be. What if a poll is done by mail and asks about whether people read their junk mail or just throw it away? Nonresponse is a reason that the number of people reported in opinion polls is often not a nice round number but something like 1,014, in which case, there were probably 486 out of 1,500 people who did not respond.

A special form of nonresponse is when a poll asks potentially embarrassing or otherwise loaded questions; in which case, people may not want to answer or may simply not tell the truth. This could, for instance, be questions about drug use or illegal behavior. A clever trick in this situation is to ask everybody to roll a die before answering the question. If the die shows 6, the person answers "yes" and otherwise the person answers truthfully. In this way the interviewer never knows if an affirmative answer is truthful or due to rolling a 6. How is the true proportion then estimated? For example, suppose that 6,000 people are polled and 3,000 answer "yes." As we expect 1,000 people to roll a 6 with the die, we expect that 1,000 of the 3,000 "yes" answers are from these rolls and the remaining 2,000 are truthful. We thus count 2,000 out of 5,000 "yes" answers, and our estimate becomes 40%. The number of 6s will of course rarely be exactly 1,000, and the effect of the randomness in the die rolls will be reflected in a wider margin of error than an ordinary poll. When this special type of poll is done, it has been decided already from the beginning that one sixth of the sample be wasted.

The political parties also do their own polling, and somehow magically they always seem to get results that support their own candidates. Other than introducing selection bias and disregarding nonresponse bias, it is also possible to introduce bias in the results by the phrasing of the questions, on purpose or unconsciously. A poll in 2005 regarding the tragic Terri Schiavo case might have been an example of unclear phrasing. In this poll, 55% sided with Terri's husband and 53% sided with her parents—an overlap even though the two parties held diametrically opposite positions. All in all, the polls done by the major polling companies are well planned and executed and give accurate results. After all, there is competition going on between the polling companies and nobody wants to face the fate of the *Literary Digest*.

Serious polls that are done by random sampling are often called "scientific." In contrast, "unscientific" polls are, for example, when people are asked to call a TV show or vote on a website. Such polls have little value other than entertainment because they suffer from *self-selection* bias. I remember one particularly unscientific poll in Sweden in 1990. It was time for the census, and unlike the United States, Sweden does not have the census mandated in its constitution and there can be quite a bit of resistance against the government intruding in people's lives by requiring them to fill out the census form (Swedes are otherwise fairly tolerant of governmental intrusion). On an evening talk show with the charismatic and entertaining host Robert Aschberg (hard to describe to a non-Swedish audience), the census was discussed and Aschberg pulled out his form and set it on fire. At the same time, people were asked to call in for or against the census and, lo and behold, 95% of the callers were against! Certainly no "delicate handling" of statistics, but it was fun to watch. By the way, the response rate for the census ended up being 97.5%, so if this had been a serious opinion poll, it would by far have beaten the *Literary Digest* for the all time low.

THE LAWSUIT AND THE LURKER

In the 1970s a lawsuit was brought against the University of California at Berkeley accusing the university of discrimination against female applicants in its graduate programs. The basis of the lawsuit was data of the percentages of males and females that were admitted to graduate school. In the six largest majors, 44.5% of male applicants were accepted versus only 30.4% of female applicants. As 44.5% is quite a bit more than 30.4%, the suitors believed this to be a case of discrimination. Now, Berkeley has always had a very strong statistics department, so the university's leadership decided to put the statisticians on the case. In order to find the departments that were the worst culprits in this flagrant discrimination, numbers were broken down by the students' majors. Oddly enough, once this had been done, there was no discrimination against female applicants to be found anywhere. In fact, it even looked like it was easier for women to be admitted to most departments! How was this possible?

There were over 100 majors, and to simplify matters, I have used data from the six largest majors that together accounted for over one third of all applications. Moreover, I classified majors as "easy" or "difficult" in reference to whether it was relatively easy or difficult to be admitted (not whether the actual subject was easy or difficult to study). The university did not allow departments to be identified, but one may imagine that science and engineering departments have the resources to admit more students than cultural anthropology or comparative philology. And after doing this, I see the exact same pattern that the statisticians did in 1973: In each of the categories, a higher percentage of women were admitted, but overall, a higher percentage of men were admitted. This sounds absurd, but look at Table 8.4, where the number of admitted students and the total number of applicants is given in each category as well as the overall numbers.

Table 8.4 Admission numbers for females and males in different categories of majors

	Female	Male
Easy major	106 of 133	864 of 1,385
Difficult major	451 of 1,702	334 of 1,306
Overall	557 of 1,835	1,198 of 2,691

The numbers given in Table 8.4 are not very illustrative, so let us convert these numbers to rates (percentages):

Table 8.5 Admission rates for females and males in the two categories of majors

	Female	Male
Easy major	79.7%	62.4%
Difficult major	26.5%	25.6%
Overall	30.4%	44.5%

Table 8.5 shows that a significantly higher percentage of women were admitted in the easy majors and that there is essentially no difference in the difficult majors. Still, overall, a significantly higher percentage of male applicants were admitted. The numbers don't lie, but still, is there a more intuitively appealing explanation? Sure. Let us look at one more statistic that has to do with what majors male and female applicants chose. In Table 8.4 we can see that the total number of male applicants was 2,691 and that about half of those chose a difficult major. The situation is different for women, where a large majority, 1,702 out of 1,835, chose a difficult major. Table 8.6 gives the proportion of men and women that applied to the two categories.

With this final analysis, the riddle is resolved. The resolution lies in considering not the *admission* rates but the *application* rates. There was no discrimination against women by the university; rather, women discriminated against themselves by applying in disproportionally large numbers for majors where it was difficult to be admitted. If women had applied in the same proportions as men and the admission rates had stayed the same in the two categories,

Table 8.6 Application rates for females and males in the two categories of majors

	Female	Male
Easy major	7.3%	51.5%
Difficult major	92.7%	48.5%

it would instead have appeared that women were favored in the admissions process.

The Berkeley admission data provide an example of *Simpson's paradox*, named after statistician E. H. Simpson, who wrote about it in the 1950s, but it had been discussed 50 years earlier by Scottish statistician G. U. Yule.[2] The choice of major, which turns out to be the real culprit, is called a *lurking variable* and might easily be missed unless you know to look for it. The Berkeley statisticians knew and got the university out of the legal mess. The reason why female applicants gravitate toward the difficult majors is of course a completely separate issue.

A hypothetical example that perhaps even more clearly illustrates the paradox is to consider high-school students who apply to college. Suppose that the students are classified as "strong" or "weak" and that we examine the proportion within each category that is accepted into their first choice of college. It would then not be surprising if a much higher proportion of the weak students are admitted to their first choice. Conclusion? Discrimination against strong students? Of course not. The strong students simply chose colleges that are harder to get into, whereas the weak students settled for easier choices. Let's make up some numbers and make them a bit extreme to illustrate the point. Consider a small high school with 10 strong and 90 weak students in the graduating class. Of the strong students, nine chose Harvard first and only one got in. One weak student also took a shot at Harvard but was not admitted. Thus, the admission rates at Harvard were 11% for strong students and 0% for weak students. The remaining students, 1 strong and 89 weak, all applied to the local community college, and all but 9 particularly weak students were

[2]Naming it 'Simpson's paradox" is an example of *Stigler's Law* that no scientifi c discovery is ever named for its discoverer, another example being when the normal distribution is called the Gaussian distribution. Incidentally, Stigler's Law was formulated by sociologist Robert Merton.

admitted, making the admission rates there 100% for strong students and 90% for weak students. As expected, strong students are more easily admitted at both places, but the overall admission rates are still 89% for the weak students and only 20% for the strong students. Just like the women at Berkeley, the strong students made it harder for themselves to be admitted by their choice of college.

In both the Berkeley data and our fictitious admission example, it makes sense to study the two cases separately rather than looking at overall rates. There are cases when the opposite applies, and one such case involves batting averages in baseball. If you compare two batters, it can happen that one has a higher batting average in each half of the season, but that the other has a higher average over the entire season (you have to check for yourself whether this has actually happened). For example, in the first half of the season, Player A hits 4 out of 10 and Player B hits 10 out of 40 for batting averages of 0.400 and 0.250, respectively. In the second half, Player A hits 5 out of 40 and Player B 1 out of 10 for batting averages of 0.125 and 0.100, respectively. Thus, Player A has a better average in each half of the season, but over the entire season, Player A has a batting average of $9/50 = 0.180$ and Player B comes out on top with $11/50 = 0.220$. Unless there is a good reason to suspect that it was easier to hit during the first half of the season when Player B was at bat more often, it probably makes more sense to look over the entire year and conclude that Player B is the slightly better batter of the two (but still not very good).

When George W. Bush was elected president and Al Gore won the popular vote in the 2000 election, this phenomenon is somewhat related to Simpson's paradox as well. If electoral votes had been allotted proportionally to the popular vote, Al Gore would have gotten more. Now, they are instead allotted to states in such a way that each state gets a number of electors equal to the number of legislators the state has in Congress (two Senators + the number of House Representatives), and the candidate who wins a state gets all electoral votes from that state. Al Gore got over half a million more votes nationwide but five fewer electoral votes. It did not help him to win by well over a million votes in both California and New York when one vote's margin in each state would have been enough. And in the end, it was a 537 vote margin in Florida that awarded George Bush the Presidency. Theoretically a candidate can win the popular vote by a large margin and yet only get 3 of the 538 electoral votes, for example, by winning Alaska by a landslide and lose every other state by a narrow margin. The newly elected president would then be more popular than his opponent in 49 of the 50 states, but the runner-up is still more popular nationwide.

On a more formal level, Simpson's paradox can be stated in terms of conditional probabilities. To illustrate this, let us reconsider the Berkeley admission example and introduce some events regarding a randomly chosen applicant:

A: Applicant is admitted

D: Applicant applies for difficult major

E: Applicant applies for easy major

and denote probabilities pertaining to female applicants by P_F and those pertaining to males by P_M. The fact from above that overall admission rates were lower for females than for males then translates into

$$P_F(A) < P_M(A)$$

When we take choice of major into account, we get the following relations between conditional probabilities:

$$P_F(\text{A given D}) > P_M(\text{A given D})$$

and

$$P_F(\text{A given E}) > P_M(\text{A given E})$$

which together illustrate that the admission rate was higher for females within each category. Now, by the law of total probability, we have, for females

$$P_F(A) = P_F(\text{A given D}) \times P_F(D) + P_F(\text{A given E}) \times P_F(E)$$

and for males

$$P_M(A) = P_M(\text{A given D}) \times P_M(D) + P_M(\text{A given E}) \times P_M(E)$$

and the resolution to the paradox lies in the probabilities $P_F(D), P_F(E), P_M(D)$, and $P_M(E)$, which describe how females and males choose majors.

We can also formulate Simpson's paradox strictly as a mathematical problem and make it lose all its charm. The question is then: Is it possible to find numbers between 0 and 1, call them A, a, B, b, p, and q, which are such that

$$A > a \text{ and } B > b$$

even though

$$p \times A + (1 - p) \times B \; < \; q \times a + (1 - q) \times b\,?$$

No problems here. Just let $A > a > B > b$, and then choose p close enough to 0 and q close enough to 1 such that the last inequality holds. Ask your mathematician friends this question, and then ask them if they believe that it is possible that women have higher admission rates in each of two categories of majors, yet a lower overall admission rate. Although these are actually variants of the exact same question, it would not surprise me if you get the answers "of course" and "no way."

FOOTBALL PLAYERS AND GEYSER ERUPTIONS

On page 154, I mentioned Sir Francis Galton and his concern that properties of eminence tend to diminish in the progeny of eminent men. We will now take a closer look at what type of data Sir Francis analyzed and what he observed in the observations I mentioned earlier, the heights of fathers and sons.[3] If you measure the heights of a large number of men and the heights of their sons, you get a number of pairs of heights (x, y), where x is the father's height and y the son's height (if there are many sons in the family, the same value of x will simply be repeated as many times as needed, paired with different values of y). In brief, what Galton observed was that the taller the father, the taller the sons, but this relationship was however not perfect. The heights of the sons experienced individual variation and on average tended to be somewhere between the height of the father and the overall average population height. As Galton was interested in properties of eminence and apparently counted tallness among them, he noted that this particular property tended to be less pronounced in the sons and thus tended to "regress toward

[3]Sir Francis' first observations of the phenomenon were actually in the sizes of garden peas, but it is unclear to which extent the loss of horticultural eminence caused him any sleepless nights. On humans, Galton measured several variables other than height, for example, "weight in ordinary indoor clothes," "strength of pull as archer with bow," and "swiftness of blow."

mediocrity" as generations passed. Continuing this line of thought seems to lead to the conclusion that the entire population would end up being of average height, but this is not what happens in reality because there is a lot of random variation to counteract the regression effect.

Let us look at the type of data Galton gathered. In Figure 8.1, there are 500 observations of heights of fathers and sons. These are not Galton's original data but instead computer generated by your humble author. Still, they are representative of what Galton observed.[4] The heights of the fathers are on the x axis (horizontal axis), and the heights of the sons on the y axis. The average height of both fathers and sons is 68 inches, which is in the center of the cloud of points. Note how the cloud is slanted upward and to the right; this indicates that taller fathers tend to have taller sons. The relation is not perfect though, and if you look at fathers' heights around 72 inches, quite a bit above average, you notice that the heights of sons vary between 68 and 74.

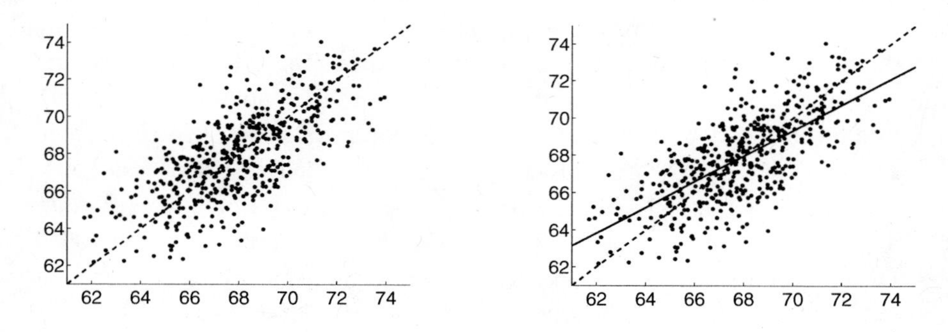

Figure 8.1 Heights of fathers on the x axis and sons on the y axis. The regression line is solid, and the diagonal line is dashed.

What worried Galton is what you can observe in the left plot in Figure 8.1 where a diagonal dashed line has been drawn through the cloud of points. Look at very tall fathers with heights between 72 and 74 inches, and note that almost all of their sons' heights are below the line. Thus, these eminently tall men had offspring who were, disappointingly, typically shorter than themselves. At the other end of the graph where fathers are far below average, almost all

[4]I have generated the heights with several decimals, more than would normally be measured. The reason is to make the points separate from each other. If heights were in integer inches, there would, for example, be plenty of points stacked on top of each other at $(68, 68)$. This can be indicated by stating the number of points in each position in the plot, but I wanted to avoid cluttering the graph.

sons are instead above the line, but it seems that Galton took no solace in this. Anyway, as it turns out there is yet another line to be drawn that better describes what is going on. This line, the so-called *regression line*, gives the average height of the son of a father of a given height. In the right plot in Figure 8.1, the cloud of points is once more displayed together with both the diagonal line and the regression line. Note how the regression line unlike the diagonal line always has roughly the same number of points vertically above and below. For any given height of the father, the mean height of the son can be computed. For example, if the father is 72 inches tall, the mean height of his sons is 70.8 inches, which is below the father's height but above the average of the general population. The formula for the regression line is a bit too technical to go into in this book, but both mathematical computer software such as Matlab and pocket calculators with statistical graphic functions will draw the regression line for you in a jiffy.

The phenomenon that Galton observed is thus perfectly normal and does not lead to a uniformization of heights and a population where eventually everybody is exactly 68 inches tall. In fact, the distributions of heights of fathers and sons are very similar, centered around the average of about 68 inches and with the same variability (same standard deviation of about 2.5). Tall men tend to get sons that are taller than the average population, but occasionally average or short men also get sons that are tall. Just like I mentioned in our first encounter with regression to the mean, think of consecutive die rolls. After a 6, you expect to get something lower in the next roll; and after a 1, you expect something higher, but you still don't expect to end up rolling 3s and 4s all the time. The analogy between heights of fathers and sons and die rolls is somewhat shaky because the height of a son depends on the height of the father, but consecutive die rolls are completely independent. However, the phenomenon of regression to the mean is observable in either case.

Regression to the mean shows up in a variety of situations, a typical one being test scores. Suppose that Carol takes two equally difficult tests, and being a good student, she can expect to score 80% on each. If she scores below 80% on the first, she is likely to improve her score on the second test. Conversely, if she scores above 80%, she is likely to lower her score on the second test. If she is lucky enough to get a really bad score on the first test, she can show a lot of improvement on the second! In contrast, those poor sods who did very well on the first test can expect their scores to drop on the second. As I have pointed out, there is a tendency among us humans to look for explanations even when none are needed. If Carol drank a lot of coffee between the tests, that may well be used to explain why her scores improved.

If schools are being rewarded for improvement as is done in some U.S. states, regression to the mean tells us that a poorly performing school is off to a much better start and is more likely to be rewarded than a good school. The same phenomenon arises in sports, medicine, and a range of other situations where it is easier to improve if you start out in bad shape and vice versa. It is a well-known phenomenon in baseball that the rookie of the year typically lowers his batting average in his second year (the "sophomore slump"), and this can easily be explained by regression to the mean. His rookie year's batting was atypically high; this deviation made him rookie of the year and is not likely to be repeated the next season. The lesson to be learned here is that it is important not to confuse real changes with the perfectly normal fluctuations that cause regression to the mean.

In modern statistics, Galton's original meaning of regression as something that strives backward has been long abandoned, but the term "regression" lives on. These days it is used to describe any kind of relation between two or more sets of variables. The simplest form is *linear* regression, which is what we did above: fitting a straight line to a dataset, but there is a multitude of other types of regression, for example, quadratic, cubic, logarithmic, logistic, and multiple, which describe different types of curves and functions fitted to the data.

When Galton assembled his datasets and discovered regression, he also investigated what came to be known as *correlation*, which is a measure of the degree of relationship between two variables. Two variables are said to be correlated if, loosely speaking, values of one can be used to predict values of the other. The height's of fathers and sons above is a good example; if you know the height of the father, you can predict the average height of his sons, and the plots we have seen illustrate the relationship. The two variables do not have to be of the same type, such as height–height. In Figure 8.2, the heights and weights of the players on the 2006 roster of the Minnesota Vikings have been plotted, and as expected, there is a clear correlation between the two. Taller players tend to be heavier, but the relationship is not perfect. As you already know what the regression line means, I have drawn this line as well. Note how the points are scattered along horizontal and vertical lines; this is because weights and heights are measured in integer pounds and inches, so there are many players that have the same weights and many players that have

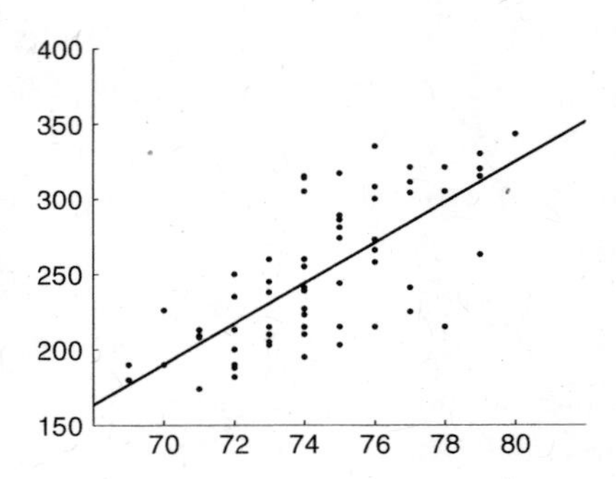

Figure 8.2 Heights in inches and weights in pounds of Minnesota Vikings players.

the same heights. With more refined measurements, we would not see these lines.

A perhaps more interesting example is provided by the Old Faithful geyser that was mentioned on page 136. When you visit this majestic geyser in Yellowstone National Park, you will notice that an estimated time for the next eruption is posted in the visitor's center. This estimate is based on an observed correlation between lengths of eruptions and times between consecutive eruptions, where longer eruptions correspond to longer times between eruptions. This makes sense because a long eruption discharges more heat and water, which means that more time is required for the pressure to build up to the point of a new eruption. Thus, once an eruption is over and its length has been recorded, this value can be plugged into the formula for the regression line to predict the time for the next eruption. Plenty of other factors influence the intereruption times, so the relation is not perfect, but it is clearly there. Figure 8.3 shows a plot of 20 eruption length–intereruption time pairs and the regression line based on these values. Old Faithful is probably the world's most studied geyser, and data have been collected for decades, so there are tons of observations to use for the predictions. I just used these 20 observations to illustrate the type of relationship on which the predictions are based.

Correlation can be quantified by the *correlation coefficient*, which is a number between -1 and 1. A correlation coefficient near 0 means that there is no linear relation between the two variables, and a correlation coefficient near -1 or 1 means that there is a very strong relation. There is a formula for how to compute the correlation coefficient from a set of observations, but as it is somewhat complicated and does not provide much insight into what correlation means, I chose not to include it here. In both the father–son data

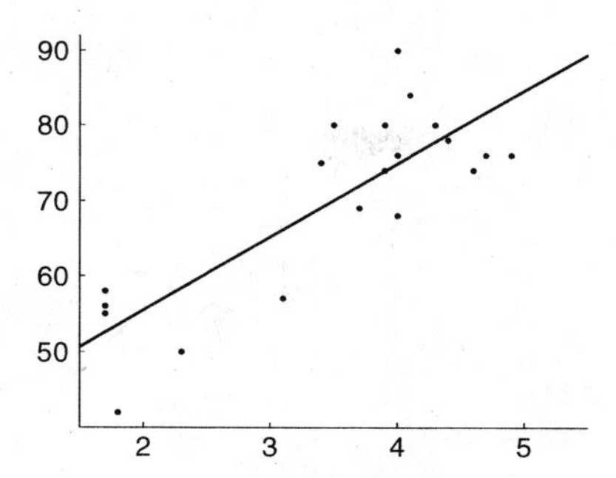

Figure 8.3 Eruption lengths in minutes and times until the next eruption in minutes for Old Faithful.

and the football player data, the correlation coefficient is about 0.7, and for the Old Faithful data, it is about 0.8, which are values that indicate a pretty high correlation (and one that is easily understood). When the regression line has an upward slope, we say that there is a *positive* correlation. A negative value of the correlation coefficient means that the regression line has a downward slope; higher values of one variable correspond to lower values of the other, and we say that there is a *negative* correlation (one example could be blood pressure and life expectancy).

If the correlation coefficient between two variables is 0, the variables are said to be *uncorrelated.* Figure 8.4 shows a plot of birthdates and weights of Minnesota Vikings players. There is no discernible pattern in the cloud of points, and of course there should not be. Birthdate and weight are uncorrelated variables. One practical comment is that when the correlation coefficient is computed from a dataset of uncorrelated variables, it will almost never be *exactly* 0, but we cannot conclude that variables are correlated unless the correlation coefficient is quite a bit away from 0 (just like in the opinion polls, we need to deal with margins of error).

Correlation provides an excellent example of Galton's advice that statistics must be "delicately handled" and "warily interpreted." One common mistake is to confuse correlation with *causation*, to believe that if two variables are correlated, one must cause the other. An amusing example of this fallacy that is sometimes seen in the statistics literature is that a positive correlation can be observed between the number of firefighters sent to a fire and the size of the economic damages to the property from the fire. Conclusion? If your house is on fire, keep the economic losses down by not calling the fire brigade? But

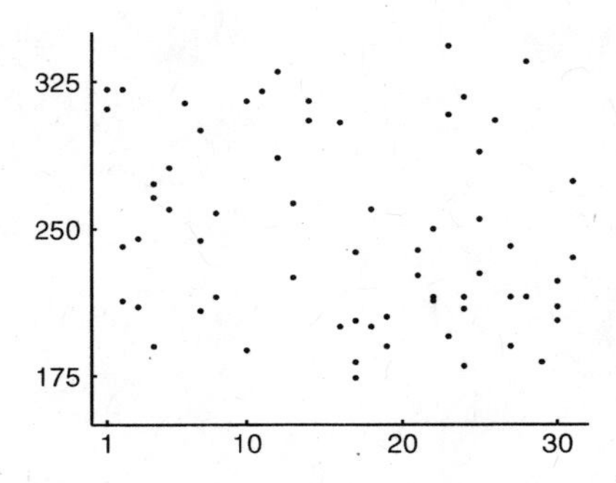

Figure 8.4 Birthdates and weights in pounds of Minnesota Vikings players. The two variables are uncorrelated.

of course there is no causation here, merely a correlation due to a lurking variable: the size of the fire. Larger fires cause both more firefighters to arrive at the scene and more economic damage. In a way, this is similar to the Berkeley admissions data from the previous section. There, we could observe a correlation between gender and admission rates, and the lurking variable was intermediate rather than underlying. Gender affected choice of major, which in turn affected the chance of being admitted.

There are many similar examples of so-called *spurious* correlations. For example, most American families would experience a positive correlation between the amount of money spent on ice cream and the size of their electric bills. Do increased ice cream purchases cause more stress on the freezer? Do increased electric bills cause more stress on the family members, making them seek comfort in ice cream? Probably not. The lurking variable in this case is, in a word, summer. Hotter, more air condition, more ice cream. At the same time, there is less need for heating in the summer, so gas bills are lower and thus negatively correlated with ice cream purchases. In these cases, there are logical explanations of why the correlations appear, but in other cases, they can show up just by chance. As your ice cream purchases probably go up every summer, they will be correlated with anything that goes up (or down) in the summer. Perhaps one year the Dow stock index went up or the number of games won by the New York Yankees went down, all mysteriously associated with your visits to Ben and Jerry's.

SNOOPING IN THE ABBOT'S GARDEN

I have mentioned that statistics is quantified common sense, and it is important to let the quantitative aspect and the common sense work together. For example, suppose that a new drug is being tested for a serious disease that, if left to run its course untreated, has a recovery rate of 80%. Now suppose that a group of 100 patients is given the new drug and that 90 of them recover, a recovery rate of 90% in this group. Can we conclude that the medication improves recovery rate?

Just like in opinion polls, we are interested in these 100 patients as a representative sample of everybody who could possible get the disease, so the mere fact that 90 is more than 80 is not enough because there is natural random variation around the expected 80% recovery rate. For example, if 82 patients had recovered, this number would be close enough to the expected 80 that we would consider it natural and not attribute it to the medication. But what about 90? Is this likely to be within the range of natural variation? If we assume the medication has no effect, the probability that 90 or more patients in a group of 100 recover can be computed to be about 0.002 or 2 in 1,000. This is pretty small, and we may feel pretty certain that the improvement was indeed because of the medicine. We have quantified the risk of this being the wrong conclusion: In 2 out of 1,000 cases, we would see such a large improvement even if the drug were useless. The numbers speak strongly for the medication, but common sense (in this case, medical expertise) still plays a role because there was of course a reason why this particular drug was tested in the first place. If a similar improvement had been shown among patients who had been given a combination of sugar pills and aromatherapy, the numbers would have been the same but not the conclusion. Such a situation would have been reminiscent of the spurious correlations from the previous section, and thus shows that in order for conclusions to be really meaningful, they need to be based on a combination of numbers and knowledge.

Here is an example that illustrates a serious misuse of statistics. I went to the Texas Lottery website and looked at the number frequencies for the game "Pick 3," where you pick three numbers between 0 and 9. At the time I looked there had been 3,822 numbers drawn in the day drawings over a period of about five years, and we would thus expect about 382 of each of the numbers 0–9. Of course we don't expect to get exactly 382 of each; some random variation is expected, and there are numbers drawn 366 or 390 times, which is perfectly normal. However, when I looked at the number 3, I noticed that it had only been drawn 345 times, which seemed to be a little on the low side. A computation reveals that the probability to get 3 no more than 345 times when you draw 3,822 numbers at random is only about 2.3%. If this had been an opinion poll where 345 out of 3,822 had expressed support for a political candidate, the usual margin of error had not been wide enough to include the expected 10% frequency and we could thus say that the deviation is statistically significant. Should we start looking for reasons why 3 is underrepresented?

No. This was a case of *data snooping*, the perfidious practice of first searching the data for anomalies and then doing calculations to prove that these anomalies are real. What is the problem with this? Well, chances are that you will find *something* that is wrong (or as statisticians joke, if you torture data long enough, it will confess). In the case of the Texas Lottery data, it may be true that there is only a little over a 2% chance to get as few 3s as we had, but this is not relevant because there is a much higher chance that *some* number is either under- or overrepresented to this degree. When I instead looked at the night drawings, I found that the number 6 was overrepresented in a way that had about a 2.5% probability. There will always be some number that deviates a little more than the others, and this is perfectly natural. Don't forget common sense; unless you are a rabid numerologist, there is no logical reason why the number 3 would be underrepresented, and if somebody claims that it is, you should be very skeptical even if statistical evidence is provided (or maybe *especially* if statistical evidence is provided).

For the lottery data, there is a clever test called a *chi-square test* that can be done to check whether the numbers are randomly drawn. The basic idea is to compute a single number that is a measure of how much the observations deviate from what is expected by randomly drawn numbers. If this single number is too large (where "too large" is quantified by probabilities), the conclusion is that something is wrong with the drawing procedure and that numbers are not random. In Table 8.7, observed and expected frequencies for the numbers 0–9 are listed. Are the deviations between observed and expected frequencies normal, or do they indicate that there are some serious flaws in the drawing procedure?

The chi-square test is based on a statistic that is calculated by first computing the square of each of the ten differences between the observed and the expected frequencies, then dividing each such square by the expected frequency, and finally adding all these numbers. If the resulting sum is large, this indicates large deviations between the observed and the expected. I will not go into detail why the statistic is computed in exactly this way, but you can

Table 8.7 Observed and expected frequencies of the numbers 0–9 in the Texas Lottery game 'Pick 3''

Number	0	1	2	3	4	5	6	7	8	9
Observed	366	382	377	345	386	390	371	412	419	374
Expected	382	382	382	382	382	382	382	382	382	382
Difference	−16	0	−5	−37	4	8	−9	30	37	−8

note that it makes sense to square the differences because otherwise negative and positive deviations would cancel each other. It also makes sense to divide by the expected number because the larger this is, the more natural random variation would we expect to see. The chi-square statistic is denoted by the Greek character χ^2, and in this case, we get

$$\begin{aligned}\chi^2 &= (366 - 382)^2/382 + (382 - 382)^2/382 + \cdots + (374 - 382)^2/382 \\ &\approx 10.95\end{aligned}$$

and the question is whether this is a large or a small number. One first comparison can be done of the chi-square statistic with its expected value. It can be shown hat the expected value equals the number of *classes* minus one, in this case ten (classes 0–9) minus one, so the expected value here is 9. The observed value of 10.95 cannot be considered to be far from the expected value especially as it can also be shown that the standard deviation is the square root of twice the expected value, here $\sqrt{18} \approx 4.24$. As 10.95 is well within a standard deviation of the expected value, the observed value is not in any way extreme, which can be further quantified by computing the probability to get a χ^2 value that is at least this large for a set of true randomly drawn numbers. With a little help from Matlab, this probability can be computed to be about 28%. Now, this is not a small probability, so our observed χ^2 value of 10.95 is not an unusually large number. The conclusion is thus that the deviations are perfectly normal, and there is no reason to suspect that the numbers are not drawn randomly. Notice the difference between this conclusion based on the entire dataset and the conclusion that the number 3 is underrepresented when it is singled out. The χ^2 test is the right approach in this situation.

The lottery data illustrate a general observation that if you have many different classes (in that case the ten numbers 0–9), chances are that you can find one that deviates more than it is expected to. However, if the chi-square test does not indicate anything abnormal, there is no immediate reason to pay any attention to the individual deviation. Another good example is to consider the 50 U.S. states. If you, for example, consider some disease or cause of death that has no reason to follow any particular geographic pattern, it is still likely that you will find some state that by pure chance deviates from the norm. Before you start looking for explanations among the particulars of the state in question, better do a chi-square test to get a more complete picture of the situation.

Data snooping is a serious problem whether it is done on purpose or accidentally. Numbers don't lie, but they may be selected to fit a particular hypothesis, and this may be difficult to reveal. Sometimes the conclusion just doesn't seem to make any sense and we may then suspect that it was generated by first looking at the data and then doing the analysis. For example, Swedish statistician Holger Rootzén once told me that he was analyzing disaster claims data from insurance companies, and noticed the peculiarity that the last claim always seemed to be the largest. What was going on here? Were disasters suddenly becoming larger? The answer turned out to be much more mundane: After an insurance company experienced an unusually large payout, they tended to seek help from statisticians. Thus, when the statisticians got their hands on the data, the last claim was typically by far the largest. And although it might have been possible to argue for sudden increased payouts based on the numbers alone, that would have missed the real reason.

A few years ago, I saw a statistical study that claimed that silver-colored cars were less likely to be involved in accidents than cars of other colors. The authors of the study offered no logical explanation, only statistical arguments, but they still suggested that more silver cars should be manufactured in order to increase security. I can't think of a reason and would suspect that silver simply popped out as being the least accident-prone color in this study. In another study, it might just as well be red or blue; just like in the Texas Lottery data, it is likely that there is some individual anomaly without this indicating anything unusual in general. In the case of the cars, there could also be logical explanations involving lurking variables such as a tendency among careful drivers to buy silver cars; in which case, the authors' suggestion to increase production of silver cars is pointless. However, since no explanation was offered a liberal dose of skepticism is recommended. But perhaps the study was just a statistical joke, which would be the most satisfying explanation.

For a classic example of data snooping in the other direction, let us consider Gregor Mendel (1822–1884), the father of modern genetics. Mendel was a brilliant man, an abbot who just like Galton experimented with garden peas. However, whereas the legumes lead Galton to despair over the loss of eminence, Mendel made far more profound discoveries and was the first to suggest that individual traits are passed on from parents to offspring as units and not blended together. In one of his experiments, he identified two different types of peas: smooth and wrinkled. When he crossed pure-bread strains of each, he noticed that all offspring were smooth, but when he subsequently crossed these offspring with each other, about 25% of peas in the next generation were wrinkled and 75% were smooth. He had discovered *dominant* and *recessive* genes. The gene for wrinkled peas is recessive, which means that it must be present in two copies to be expressed whereas one copy of the smoothness gene is enough for the pea to be smooth. If both parent plants are *heterozygotes*, that is, have one gene for each trait, they must themselves be smooth but their offspring can be either smooth or wrinkled. Call the genes S and W, so that each parent's genotype is SW. Since each parent passes on a randomly selected gene to the offspring, there is a 25% chance that the offspring receives two W and thus becomes wrinkled. With probability 75%, the offspring receives at least one S and will then be smooth. If it has genotype SW, it is a *carrier* for the wrinkled trait just like its parents. Genetics is, by the way, a scientific discipline where methods from probability and statistics are widely used.

Let us return to Mendel. In one dataset, he had 7,324 peas, 5,474 of which were smooth and 1,850 wrinkled. Under the assumptions of random mating, the expected number of smooth peas is 75% of 7,324, or 5,493 and the expected number of wrinkled peas is 1,831. The problem with Mendel's numbers is that they are *too good to be true!* Just like observed numbers can be too far from expected numbers, they can also be too close. In this case we know that Mendel's conclusions were correct, but it is commonly thought that his data were manipulated to better fit his hypothesis. Being the brilliant scientist that he was and an abbot as well, Mendel is usually raised above suspicion and it is widely believed that one of his assistants fixed the data knowing what results his master wanted.

The chi-square test can be used in other situations as well. In Chapter 3, we discussed the Poisson distribution and how it shows up in situations where rare

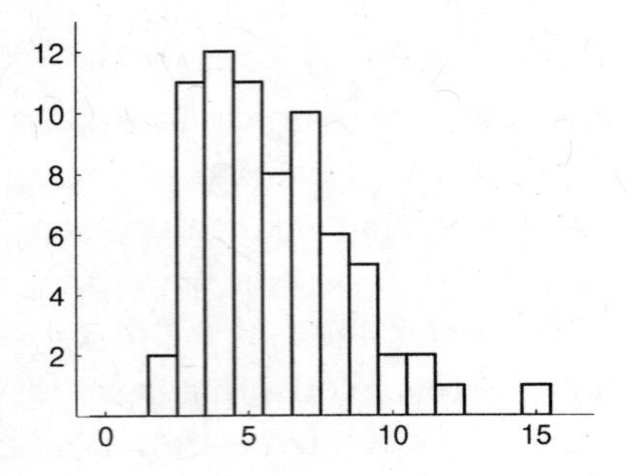

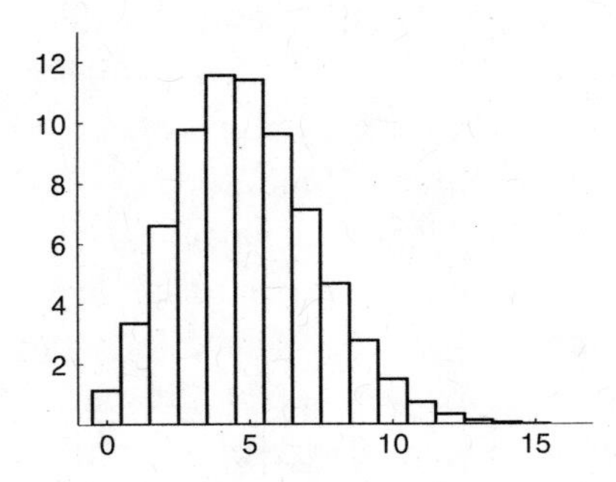

Figure 8.5 Observed and expected annual numbers of Atlantic hurricanes for the years 1935–2005. To the left are the observed numbers. There were no years with 0 hurricanes, no years with 1 hurricane, two years with 2 hurricanes, and so on up to one year (the infamous 2005) with 15 hurricanes. The right graph shows the expected number of annual hurricanes assuming a Poisson distribution.

unpredictable events are considered. As I am writing this in New Orleans and hurricane season is right around the corner, it might be interesting to examine whether the annual number of hurricanes follows a Poisson distribution. Hurricane formation is certainly unpredictable and rare on an appropriate time scale, so a Poisson distribution might be reasonable. The average number of hurricanes per year has been a little below six for the last half century, and if we assume a Poisson distribution with this expected value, we can compute the expected number of years with 0 hurricanes, 1 hurricane, 2 hurricanes, and so on and compare this with the actual numbers of years that have had 0 hurricanes, 1 hurricane, 2 hurricanes, and so on. A chi-square test reveals that the fit is good, and in Figure 8.5, you can see plots of the actual and expected frequencies for the years 1935–2005 and note that they look fairly similar.

FINAL WORD

In his book *Chance Rules*, mentioned in Chapter 2, Brian Everitt points out what I also alluded to in the beginning of the book: Statisticians are not appreciated at cocktail parties. I hope that the chapter you have just read has made you look a little kinder on the field of statistics and its practitioners. Statistics is without a doubt a discipline that is only growing in importance as we get more and more involved in gathering and analyzing data of all kinds. A somewhat unfortunate sign of the importance of statistics is the completely meaningless use of statistical terms in all kinds of contexts. I recently heard a TV commercial for a skin care product claiming that if you used it, your skin would look "on average ten years younger." I have no idea what this is supposed to mean.

Other than being useful, statistics can also be really interesting and not at all as dry and boring as a lot of people think. If knowledge was more widespread about the German tanks, the Berkeley lawsuit, Mendel's cheating assistant, and other examples that we have looked at, perhaps statisticians would not have to go to social functions pretending to be astronauts or dolphin trainers or even actuaries to avoid isolation. Maybe they would even be the center of attention, who knows? After all, a very clever man once said that statistics are full of "beauty and interest" and who could ever get enough of that?

9

Faking Probabilities: Computer Simulation

MAHOGANY DICE AND MODULAR ARITHMETIC

Simulation is one of the most commonly used techniques to gain information about complicated systems, but the term "simulation" is used to convey many different meanings. According to the Merriam-Webster online dictionary, simulation is "the imitative representation of the functioning of one system or process by means of the functioning of another." We probably think of simulation as something involving computers, but it does not have to be so. For example, airplanes flying in specific parabolic patterns are used in astronaut training to simulate weightless conditions in space. Even when we restrict our attention to computer simulation, there are many different meanings. An airline pilot in training sits in a flight simulator, a nurse in training may use a medical simulator that imitates reactions of a real patient, and the Weather Channel may run a computer simulation with graphics to illustrate how a hurricane can affect coastal areas. For probabilists, however, simulation means "imitating randomness" and the term "Monte Carlo simulation" is sometimes used to emphasize this when we talk to outsiders.

The main use of simulation is to approximate quantities that are difficult to compute exactly. For example, compare roulette and blackjack. In roulette, it is easy to compute probabilities and expected gains, but in blackjack with its more complicated rules and choice of strategies, exact probability calculations can be very difficult. It is easy to run a computer simulation of the game though and keep track of how much you win or lose in each round with some particular strategy. Let the computer run a large number of gambling rounds, and compute the average gain over all rounds; this number is approximately your expected gain. Notice "large number" and "approximately" in the last sentence. This means that we are relying on the law of averages (which you may recall is also known as the law of large numbers) to use our observed value of $\bar{X}$ to approximate the true expected gain μ, which is unknown and difficult to compute exactly. As computers are fast, we can usually guarantee a large number of runs and this is crucial to the reliability of simulation results.

The object of interest in simulation does not necessarily have anything to do with randomness itself. An early example of simulation is Buffon's needle, where π can be approximated by a needle tossing simulation. The needles themselves are of no interest; they are merely a tool to get to the value of π. A more modern and useful example is *Monte Carlo integration*, which is a way to use random numbers to compute the area under a curve, such as the one in Figure 9.1. If you knew the equation for the curve, you could find the area by methods from calculus (compute the *integral* of the function that describes the curve). If the equation is unknown, we can place the curve inside a box like is done in the right picture and then generate random points inside this box. Such a point hits the area under the curve with a probability that equals the fraction that area is of the total area of the box, so if we generate a lot of random points and count the proportion that falls under the curve, this proportion will be approximately equal to the fraction of the area that is under the curve. This is the law of averages in action telling us that a relative frequency stabilizes at the true probability. In the figure I have plotted 200 observations, 84 of which are under the curve, which gives a fraction of 0.42. The true fraction is 0.416, which is very close. In this particular case, I was probably a little luckier than I deserved, but in order to get a reliable value, I would generate hundreds of thousands of points and feel confident of obtaining the correct area with good precision.

An important use of simulation is to evaluate new and complex statistical techniques. Suppose that a statistician has developed a new method to better predict election results from exit polls. Theoretical calculations are more or less impossible, and she has no data to test her method on because it has never been used before. Computer simulation gives her an invaluable tool to test her method. She can run computer simulations of various scenarios over and over and see how close her predictions come to the true numbers (note that it is she who chooses the true numbers so she is in full control of everything).

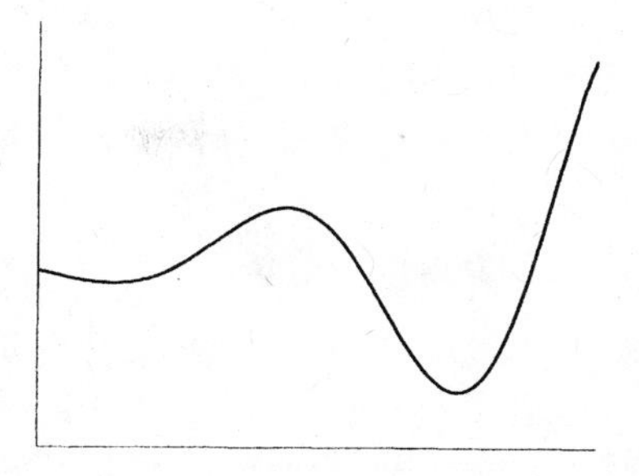

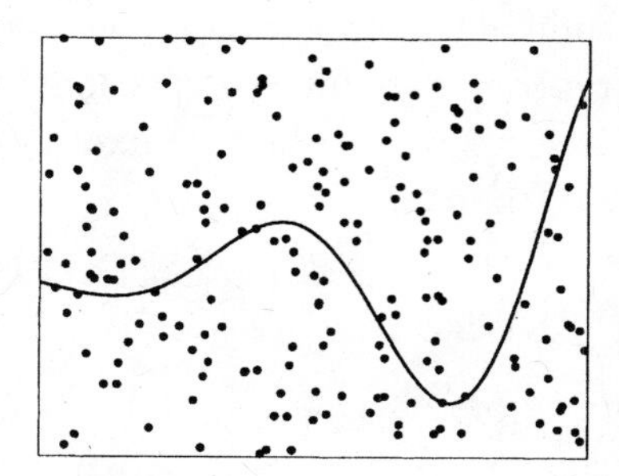

Figure 9.1 A curve (left) and the same curve in a box with 200 randomly generated points (right). The fraction of the box that is under the curve is approximated by the fraction of points that is under the curve.

The need for simulation techniques did not arise with the arrival of the computer. The usefulness of procedures to create randomness did of course not escape Sir Francis Galton, and in 1890, he wrote an article entitled, "Dice for Statistical Experiments," published in the journal *Nature*. Says Sir Francis:

> As an instrument for selecting at random, I have found nothing superior to dice. It is most tedious to shuffle cards thoroughly between each successive draw, and the method of mixing and stirring up marked balls in a bag is more tedious still. A teetotum or some form of roulette is preferable to these, but dice are better than all. When they are shaken and tossed in a basket, they hurtle so variously against one another and against the ribs of the basket-work that they tumble wildly about, and their positions at the outset afford no perceptible clue to what they will be after even a single good shake and toss.

Galton goes on to describe how his dice are constructed. They are cubes with 1 1/4-inch sides, made of mahogany (what else?), and marked on faces as well as edges to give more than six outcomes per die when their orientation is taken into account. He also constructs dice that are responsible for giving plus or minus signs and concludes that the "most effective equipment" seems to be to use six dice.

It is touching to notice Galton's concern that the dice really gave completely unpredictable outcomes. In contrast, our modern computer-generated random numbers are in reality not random at all; we are cheating big time. A computer program can't do anything as spontaneous as produce a random number. Instead it produces a sequence of numbers generated by some algorithm that computes each number from the preceding.[1] The numbers are not random; they just look random. Still, we do not have to be quite as harsh as Detective Lieutenant Ed Exley, played by Guy Pearce in the 1997 movie *L.A. Confidential*:

> A hooker cut to look like Lana Turner is still a hooker. She just looks like Lana Turner.

Far from it. The so-called *pseudo-random* numbers generated by computers are very useful even if he who knows the algorithm can predict each number with absolute certainty. Let us look at how random numbers are created. There are many different methods to do this, and I will only describe one of the most common types, the *congruential* random number generator. Such a generator uses *modular arithmetic*, and if you think that you don't know what that is, you are wrong. For example, you know perfectly well how to perform addition modulo seven; you may just not know it yet.

Modular arithmetic is ordinary arithmetic on a finite set where the largest number is again followed by the smallest, and it starts over like the numbers on the face of a clock (and I am happy that the digital watches that everybody seemed to have in the late 1970s are gone; they were detrimental to the understanding of modular arithmetic). For another example, if you number the days of the week from Monday to Sunday by 1 to 7, after 7 comes 1 again. Now suppose that it is Tuesday. What day will it be nine days later? Obviously the same as two days later: Thursday. As Tuesday is 2, you add 9 to it and would get 11, but because you have to start over after 7, you instead get 4, and 4 is Thursday. Congratulations, you just added 2 and 9 modulo 7 and got 4! A formal way of writing this is

$$2 + 9 = 4 \pmod 7$$

We also say that 4 and 11 are *congruent* modulo 7; they are simply the same under this type of addition. Any multiple of 7 added to 4 is congruent to 4, and the set of all these numbers $\{4, 11, 18, ...\}$ is called the *congruence class* of 4 modulo 7 (you can also include the negative numbers that you get by subtracting multiples of 7, that is, $-3, -10$, and so on). If you count months, you instead perform addition modulo 12; for example, $11+3 = 2 \pmod{12}$ or in other words, three months after November comes February. And if you're

[1] Incidentally, Merriam-Webster also gives the alternative definition of simulation as a "sham object."

not American (or if you are in the U.S. Navy), chances are that you count time by adding hours modulo 24; for example, $15 + 10 = 1 \pmod{24}$; in other words, ten hours after 3 P.M., it is 1 A.M.

OK, so you did know how to do modular arithmetic. How does that help you create random numbers? A congruential random number generator creates a sequence of numbers in the set $\{1, 2, ..., n\}$ by starting with some number (called the *seed*), then multiplying this by a constant and adding another constant and computing the resulting number modulo n. Formally, after you have generated the kth number x_k, the next number x_{k+1} is obtained by

$$x_{k+1} = a \times x_k + b \pmod{n}$$

where a and b are chosen in advance. Primarily we want to achieve two things for a long sequence:

(a) All numbers should show up in the same proportions

(b) It should be impossible to predict the next number

Let us look at a simple example, a die roll. Then $n = 6$, and the set is $\{1, 2, ..., 6\}$. If we choose the seed 1 and let $a = b = 1$, it is easily checked that we get the sequence 1, 2, 3, 4, 5, 6, 1, 2, ..., which satisfies (a) but certainly not (b). Let us try the seed 1 and the values $a = 3$, $b = 5$. Then the first number is $3 \times 1 + 5 = 8$, from which we subtract 6 to get 2; that is, $8 = 2 \pmod{6}$. The next few numbers that follow are

$$3 \times 2 + 5 = 11 = 5 \pmod{6}$$

$$3 \times 5 + 5 = 20 = 2 \pmod{6}$$

and from here on the sequence repeats: 2, 5, 2, 5, ..., obviously not that good either. No other numbers show up at all, and the sequence quickly becomes highly predictable. You may realize already that we cannot do very well with this method. The best we could possibly hope for is a sequence containing the numbers 1 through 6 in some random-looking order, which then repeats indefinitely. But we can't get even a sequence like 3, 3, 1, ... because if the first 3 produced another 3, then that 3 will do the same thing. We need to do something more clever.

Instead of using $n = 6$, let us use $n = 60$, divide each number by 10, and round this up to the nearest integer to get our die roll. This means that the numbers between 1 and 10 give the outcome 1, the numbers between 11 and 20 give 2, and so on. Let us take the seed 1 and $a = 11$, $b = 13$. The first number is

$$11 \times 1 + 13 = 24$$

which we divide by 10 to get 2.4, which rounded up gives the outcome 3. To continue, plug in 24 into the algorithm:

$$11 \times 24 + 13 = 277 = 37 \pmod{60}$$

which in a similar way gives the outcome 4. The sequence of simulated die rolls continues with 1, 2, 4, 5, 2, 3, which looks better than before. Note that 2 can produce 4, but it can also produce 3 because the underlying algorithm works with numbers between 1 and 60. We are making progress, but eventually, a number between 1 and 60 will repeat and the sequence of die rolls starts repeating as well. The length of the sequence before this happens is called the *period* of the random number generator, and we obviously want the period to be as long as possible.

We can can do even better by using $n = 600$ and divide by 100 or $n = 6{,}000$ and divide by 1,000 and so on. In general we take some huge number m and let $n = 6 \times m$, divide the numbers by m, and add one to the integer part to get our die rolls. I did one attempt with $n = 6{,}000{,}000$, $a = 374{,}511$, and $b = 977{,}597$ (values of a and b arbitrarily chosen by yours truly by pounding vigorously on the keyboard) and got the following first 20 die rolls:

2 5 1 4 1 4 2 1 2 3 5 5 4 6 6 3 4 2 1 4 1 3 5 5

and a more extensive check reveals that the numbers come up in their correct proportions in the long run. Still, this is probably not a very good random number generator. We have to be careful how to choose the numbers a and b because the length of the period and hence the quality of our random sequence depends on these choices. For example, if $n = 600$, it is not a very good idea to let $a = 1$ and $b = 100$ because this will give a sequence of die rolls that is just 1, 2, 3, 4, 5, 6 over and over. The mathematical discipline of *number theory* provides results that can be used to make good choices of a and b and of n. Number theory is among the purest mathematics one can find, and its practitioners devote their lives to thinking about how prime numbers behave

Table 9.1 A sequence of random numbers and the corresponding sequence of heads and tails

Random number	Coin toss
0.9501	tails
0.2311	heads
0.1068	heads
0.4860	heads
0.8913	tails
0.7621	tails
0.4565	heads
0.0185	heads
0.8214	tails
0.4447	heads
0.6154	tails

and other things that are far too profound for most of us. Amusing then that this highly theoretical branch of mathematics has proved very useful not only in simulation but also in cryptography. One of the purest of the pure was English mathematician G. H. Hardy (1877–1947) who once prophetically said

> Pure mathematics is on the whole distinctly more useful than applied.

I am sure he did not have die rolls in mind.

Most computer programming languages and pocket calculators have random number generators. You push the appropriate button, and this usually results in a decimal number between 0 and 1, such as 0.3425 or 0.9010. These numbers are generated with methods similar to what I described above and are useful as "building blocks" for any kind of simulation. As we can get a large number of decimals even on a simple calculator, we may for practical purposes suppose that we can get all numbers between 0 and 1. To simulate a sequence of die rolls, generate a sequence of random numbers and let numbers between 0 and $1/6$ give 1, numbers between $1/6$ and $2/6$ give 2, and so on. To simulate coin tosses, let numbers between 0 and 0.5 give heads and numbers between 0.5 and 1 give tails. For an example, see Table 9.1.

If you want to simulate something where not all probabilities are equal, you simply divide the interval [0,1] into parts that correspond to your probabilities. For example, if you want to simulate the number of daughters in a family with two children, this can be 0, 1, or 2, with probabilities $1/4$, $1/2$, and $1/4$, respectively, and you let random numbers between 0 and $1/4$ give 0, numbers between $1/4$ and $3/4$ give 1, and numbers between $3/4$ and 1 give 2 daughters.

Simulating more complicated probability distributions requires more complicated maneuvers. For example, it can be shown that if you have two random numbers A and B and compute the number X according to the formula

$$X = \sigma \times \sqrt{2 \log A} \times \cos(2 \times \pi \times B) + \mu$$

then X has a normal distribution with mean μ and standard deviation σ (the "cos" that appears in the formula is the trigonometric function *cosine*). Note how the pesky π appeared again, perhaps not so surprising anymore. So if we want to simulate 1,000 IQ scores with mean 100 and standard deviation 15, we generate 2,000 random numbers and use them two by two to compute 1,000 values of X. Figure 9.2 shows 1,000 IQ scores simulated in this way.

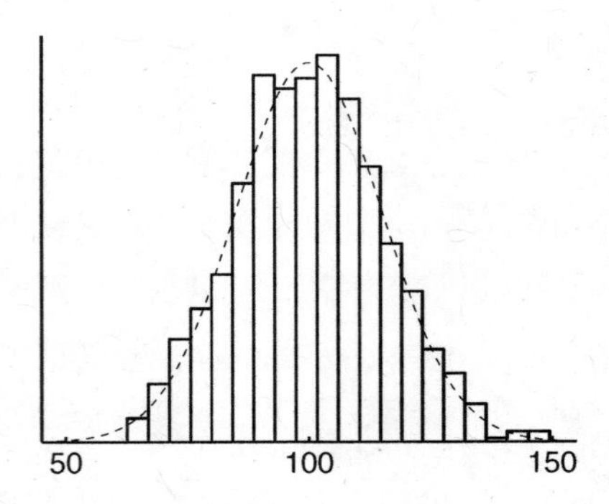

Figure 9.2 One thousand simulated IQ scores with mean 100 and standard deviation 15. The dashed curve is the theoretical bell curve.

RANDOM AND NOT-SO-RANDOM DIGITS

There was a time when people needed more extensive sets of random numbers than Galton's mahogany dice could produce but when there were yet no computers around. Instead, *random number tables* were used. These tables were created by collecting digits from various sources. One early example is a table of 41,600 random digits gathered by statistician Leonard Tippett and published in 1927. Tippett obtained his digits from census data, and he claimed that they were taken "at random" without specifying precisely what this meant. He had tried to obtain random digits by drawing numbered cards from a bag but that proved to be both cumbersome and unsatisfactory (he should have read his Galton). In 1955, the RAND Corporation published *A Million Random Digits*, which contains a million random digits (yes, really) obtained by an electronic roulette wheel. Here is an excerpt:

13073 43556 45009 13436

58884 93194 33498 01299

Or is it? Don't accuse me of plagiarism just yet; maybe I made up those numbers on my own. You can check it for yourself; the work was republished in 2001, and I have noticed that there are several used copies for sale at Amazon.com. Apparently people just memorize the numbers and get rid of the book.

On the RAND Corporation webpage, it is claimed that the table is still used, but this I doubt. Any reasonable computer software such as Matlab can produce a million random digits in a fraction of a second. Still, the table is an impressive piece of work from a time when people knew what they needed and had to be very creative to achieve it. I remember from high school that our mathematical tables also had a page with random digits. I never understood it. They were the same day in and day out. What was so random about that? With a pocket calculator you get the feeling that something random is actually produced inside the box, but as we have seen, that is an illusion. Still, I think it is nice that modern computational equipment can add a bit of mystery to life. Random number tables are so clinical.

NUMBER ONE IS NUMBER ONE

The idea to generate random digits by choosing from large datasets is a clever one, but you must be careful how to do it. Suppose that you choose the first digit each time, excluding 0 so that for example the numbers 34.509 and 0.0031 both give the first digit 3. If you try to produce a sequence of random numbers in this way, you are in for a big surprise. The numbers 1–9 will not be equally

Table 9.2 First digits and their probabilities (in %) for the populations of nations and according to Benford's law

First digit	1	2	3	4	5	6	7	8	9
Population data	28.2	17.2	13.7	10.1	10.1	5.3	6.6	6.6	2.2
Benford's law	30.1	17.6	12.5	9.7	7.9	6.7	5.8	5.1	4.6

likely! That seems like a bold and somewhat absurd statement because I don't even know where you intend to choose your numbers. However, a result known as *Benford's law* informs us that 1 is the most likely first digit with a probability of about 30% rather than the 11.1% we would get if all nine digits were equally likely. The second most common first digit is 2, followed by 3, 4, and so on up to 9, which is the first digit in only 4.6% of the cases.

Benford's law is not a mathematical theorem that can be proved. However, the law has been extensively tested and empirically observed, and in Table 9.2, I report my investigation into the first digit in the populations of the world's nations, obtained from www.nationsonline.org. Figure 9.3 gives graphs of the probabilities. The agreement between the observed probabilities and those predicted by Benford's law is remarkable. Is there a way to understand this phenomenon? Why is 1 so common and 9 so rare as first digit of population numbers?

Populations of nations typically increase at a fairly steady rate. One nation in the "nine category" is Sweden whose population recently passed 9 mil-

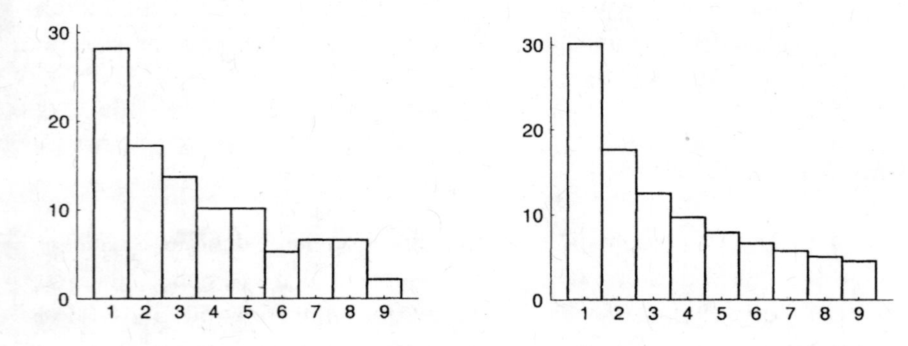

Figure 9.3 A comparison of the probabilities of first digits for the populations of nations (left) and according to Benford's law (right).

lion. Eventually it is likely to exceed 10 million and then the first digit in its population number changes from 9 to 1 and will remain 1 for a long time. Not until the population has doubled to 20 million does it change. Examples of countries whose populations numbers start with 1 are China, Japan, and Russia (such nations are much easier to find obviously), and they will stay there for quite some time. When the first digit has just changed to 9, changing it further to 1 requires an 11% population increase, whereas the next change from 1 to 2 requires a doubling in population size. Next, the change from 2 to 3 requires a 50% increase and so on and so forth. The passing from one digit to the next requires an ever decreasing rate of population change and that is why population sizes growing at a constant rate stay the longest in the lowest category and the shortest in the highest. As populations typically grow at a geometric rate (population size doubles at constant time intervals), the mathematically savvy may suspect that logarithms are somehow involved in the Benford proportions, and as we will see later, this is absolutely correct.

The explanation for population numbers hinges on growing populations and will not apply to other datasets where Benford's law is still observable. However, there is a general observation that 1 always is "ahead" as a first digit that you can make simply by counting. Start by 1, 2,..., 9, and there they are all in equal proportions. Then comes 10, 11,..., 19 and 1 is way ahead. Next 20, 21,..., 29 and 2 has caught up and so on and so forth; the digit 9 is the last to equalize the score at 99 and then comes 100 and 1 sets off on a 100-digit-long increase in its numbers. And so it goes, with 1 constantly pulling away from the field, 9 constantly lagging behind. It is of course easy to find datasets where the law fails, for example, people's heights in inches, where 6 and 7 would be the most common first digits or zip codes in California where 9 gets a total revenge. However, any dataset that is large and in some sense irregular is likely to follow Benford's law.

The law is named for Frank Benford who wrote about it in a 1938 paper published in the *Proceedings of the American Philosophical Society*, but others had made similar observations before him. Astronomer Simon Newcomb wrote about it already in 1881, and according to legend, he discovered it when he noticed that the pages in a book of logarithm tables were unevenly worn. The first pages where logarithms could be looked up for numbers starting with 1 seemed to have been used much more than later pages. I wonder if one can see a similar wear-and-tear effect on a computer keyboard. Mine looks evenly worn, but then again, I don't compute logarithms very often.

It has been suggested that Benford's law can be used to detect fraud in accounting, insurance, and elections. The idea is that if numbers are made up, chances are that the perpetrators fail to get the first digits in their correct Benford proportions, and this can be discovered in a thorough analysis. This idea was also used by Colin Bruce in his 2001 book *Conned Again, Watson: Cautionary Tales of Logic, Math, and Probability*, where he lets Sherlock Holmes solve various crime problems by applying mathematical reasoning.

You can also use the law to make up favorable bets. Choose the numbers 1, 2, and 3 as yours, and let Uncle Sid get the numbers 4–9. You then ask him to choose his favorite large dataset and choose a first digit in some arbitrary way. Whoever gets a number in his range wins. And although you only have three out of nine numbers, the Benford probabilities of 1, 2, and 3 add up to a little over 60%, so Uncle Sid loses again.

There are more sophisticated explanations of why Benford's law is valid, and these include considerations of *scale invariance*, *base invariance*, and *logarithmic distributions*. I will not discuss these here, but in order to get the exact probabilities, some mathematical assumptions that formalize why first digits behave like they do must be made. When this is done, the exact probabilities in Benford's law are computed according to the formula

$$\text{P(first digit is } d) = \log_{10}(1 + 1/d)$$

where $\log_{10}$ is the ordinary base-10 logarithm and d ranges from 1 to 9. For example, $d = 1$ gives $\log_{10}(2)$, which equals 0.30; if you raise 10 to the power 0.30, the result is 2. There is also a formula for the *second* digit that is more complicated and shows that second digits are not uniformly distributed either. The most likely second digit is 0, with a probability of 12% and the probabilities decrease down to 8.5% for 9. Again, 9 is the least likely, but the declining effect is not at all as dramatic as for the first digit. When it comes to the third digit, this declining effect has essentially disappeared.

IS RANDOM REALLY RANDOM?

Take a look at the two sequences of 30 filled and unfilled circles in Figure 9.4. One of them is random, that is, imitating a coin toss sequence where unfilled circles correspond to heads and filled circles correspond to tails. The other sequence is manipulated in some way. Can you tell which is which?

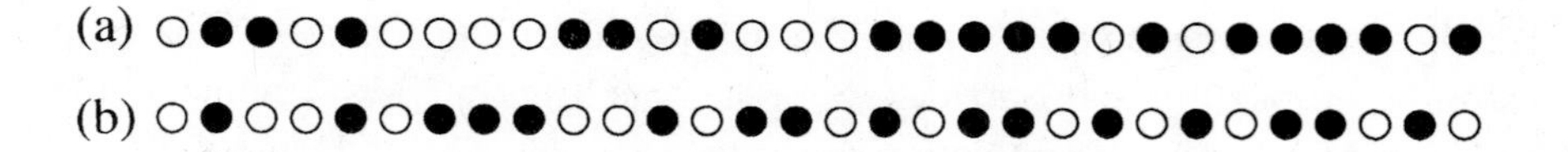

Figure 9.4 Two sequences of heads (unfilled circles) and tails (filled circles). One sequence is random, and the other is not.

This is of course impossible to tell. In true coin tosses, any sequence of heads and tails is possible and all sequences have the same probability $(1/2)^{30}$. The question is which sequence is most likely to be the one generated at random, the one displaying typical properties of a random sequence. What about proportions of heads and tails? The first sequence has 13 heads, and the second 14, so this offers no help; both sequences have their proportions close enough to 15 heads and 15 tails to pass this test. Sequence (b) does look a little better mixed though, whereas sequence (a) tends to have some suspiciously long sequences of consecutive heads or tails. Hands up now, how many of you think that this is the argument that sequence (a) is fake and (b) is real?

Take your hands down; of course it is the other way around! Sequence (a) is indeed a real coin toss sequence, and sequence (b) is fake. One aspect other than proportions that we can examine is the number of *changes* from heads to tails or vice versa. After the first toss, there are 29 possibilities to change and each time a change occurs with probability 0.5. The expected number of changes is therefore 14.5. Sequence (a) has 15 changes and sequence (b) has 22, and this indicates that sequence (b) is somehow manipulated to increase the number of changes. How extreme is it to have 22 changes? A calculation involving the binomial distribution reveals that the probability to have 22 changes or more is only 0.12%; that is, just barely more than 1 out of every 1,000 sequences will have this many changes. And now I confess: Sequence (b) was created by letting a change occur with probability 0.7 instead of 0.5. The expected number of changes is then equal to 20.3 and 22 is in no way extreme.

Note that sequence (b) still has the numbers of heads and tails in the correct 50–50 proportion. It just changes more often than sequence (a). Sequence (b) is an example of a *Markov chain*, a sequence where the next character is not chosen randomly but depends on what the last character was (but independent of previous characters).[2] Markov chains provide one of the most important tools in probability and statistics, both to model real-world phenomena and as an aid in computer simulation (so-called *Markov Chain Monte Carlo* or MCMC). We could also produce a sequence that tends to change less often than independent coin tosses, for example, by letting the probability of change be 0.3. This would result in longer sequences of consecutive heads and consecutive tails, but in the long run, they would still show up in equal proportions. The study of Markov chains is an absolutely fascinating topic, but it requires more mathematics than I care to introduce in this book.

The example illustrates that when a random number generator is evaluated, it is far from enough to demand that the digits or characters show up in the correct proportions; this is just a first necessary requirement. Sequences are also tested for so-called *runs*, that is, uninterrupted sequences of the same digit or character. In a coin toss sequence, we expect to see a certain number of runs of tails of length two (such as TT in HTTH), three (such as TTT in HTTTH), and so on. One problem with sequence (b) above is that it has too many runs of tails of lengths one and two, only one of length three, and none longer. When you generate a random sequence of coin tosses or toss a real coin 30 times, it is much more likely than not to get a run of length four. For really long sequences of coin tosses, we expect to see longer runs as well. Recall for example that we have to wait on average about a million tosses for a sequence of 20 heads, so in several million tosses, this is likely to show up.

People are notoriously bad random number generators. We can do short sequences alright, but when it comes to longer sequences, we typically tend to come up with something like sequence (b) above: too many changes. The problem is that we remember what we have done, and after we have dared to put four heads in a row, we probably think it's time for tails. And even if we are aware of the independence of coin tosses, we will simply be too concerned with getting the correct proportions. Remember how we discussed Tom and Harry's coin tossing games and how one player tended to be ahead all or most of the time even though the game was perfectly fair. We humans are not good

[2]Markov chains are named after Russian mathematician Andrey Andreyevich Markov (1856–1922) who was a student of our old friend Chebyshev. One imaginative application Markov found for his chains was for the sequence of vowels and consonants in Alexander Pushkin's 1833 poem *Eugene Onegin*. Markov empirically verified that a vowel was followed by a consonant 87% of the time and that a consonant was followed by a vowel 66% of the time. Sadly, our data-obsessed old friend Sir Francis Galton probably never heard of Markov's investigation.

at recognizing randomness and might need to remember the rule of thumb: If it looks random, it ain't, and if it doesn't, it is.

A question that mathematicians have studied is whether the digits of π are random. This may seem like an absurd question. After all, π is the ratio of the circumference to the diameter of a circle and has not changed since the days of Archimedes.[3] But just like my high-school table of random digits, the question is of course whether the digits of π look like they could have been generated at random (mathematicians call such numbers *normal*). Are there deviations from the correct proportions of the digits 0–9 or any discernible patterns?[4] The answer seems to be "no," and when the digits of π are tested against digits produced by commercial random number generators, the ancient number does pretty well. So if you happen to be a *piphilologist* and are asked to produce a sequence of random digits, just start with 3, 1, 4, and keep going. If you are afraid that your bluff will be caught, you can also memorize the square root of two or the logarithm of two and perhaps mix the digits from all three. It's up to you.

If you want to generate a sequence of coin tosses from the digits of π, you can use its *binary* representation. If you are not familiar with this concept or have forgotten about it, the binary representation is when base 2 is used instead of the ordinary base 10. For example, π starts with 3.1415, where 3 tells us that we have 3 ones, 1 tells us that we have one tenth, 4 tells us that we have 4 hundredths, and so on. In the binary system, we instead count powers of 2. As 3 equals $2+1$, the integer part of 3 on binary form is 11. Next, can we add one half? No, that would be too much so there are 0 halves. Can we add one fourth? No. One eighth? Yes, as this equals 0.125, which is less than the 0.14 found in π. When we have added one eighth, can we add one sixteenth? No, because that would land us at 3.1875, which is too much. Taken together, we have argued that π in its binary form starts with 11.0010. We can now let 1 represent heads and 0 represent tails and get a sequence of coin tosses. Here

[3]One attempt to change it was, however, made by the Indiana State legislature in 1897. The Indiana House then passed a bill asserting that the value of π was incorrect and should instead be 3.2 (some other numbers appear in the bill as well). The "Indiana Pi Bill" was postponed indefinitely in the State Senate, and it is unclear whether it will be up for a vote anytime soon.

[4]For biblical numerologists, the pattern 666 shows up for the first time after 2,240 digits. As the expected waiting time for 666 in a random sequence is 1,110, this might indicate that π is far from diabolic.

is one based on the first 20 digits of π in its binary form:

H H T T H T T H T T T T H H H H H H T H

which looks just fine, don't you think?

Although we may not always be very good at recognizing randomness, we at least understand what a randomly produced sequence of coin tosses should look like. Heads and tails in roughly equal proportions, runs and other patterns in their correct proportions, and so on. A different and more practically oriented question is how random is a *real* coin toss. On page 145, I mentioned Stanford professor Persi Diaconis whose carefully calibrated thumb can produce completely nonrandom coin tosses. His coin tossing skills combined with his probabilistic knowledge made him interested in the question of really how random a sequence of real coin tosses is. In a sense, a coin toss is of course not random at all because if we knew the initial speed and rate of spinning, the final position of the coin when it lands in your hand can be calculated with Newton's laws (the situation becomes far more complicated if the coin lands on a hard surface and bounces and spins before stopping). For any combination of initial speed and rate of spinning, it can be calculated whether the coin will land with the same face up as it started. The perceived randomness is in the uncertainty of those initial conditions that most of us with normal insensitive thumbs cannot perfectly replicate and very small changes can lead to large differences in the final outcome.[5]

In a 1986 article in *Statistical Science* entitled, "A conversation with Persi Diaconis," the magician turned probabilist explains how he used a stroboscopic light to measure initial speed and spinning rate of a coin tossed repeatedly in the air to a height of about a foot. His investigation lead him to conclude that consecutive coin tosses are not completely random, and this is due to the fact that the number of spins before the coin lands does not vary very much. In a recent article entitled, "Dynamical Bias in the Coin Toss," Diaconis and coauthors Susan Holmes and Richard Montgomery have investigated coin tosses in more detail, both theoretically by the laws of physics and experimentally using a coin tossing machine (one wonders if Sir Francis didn't have one of these hidden in his attic). Their conclusions are that a normal coin is slightly

[5]It is interesting to note that Pierre-Simon Laplace who did pioneering research in probability was also a firm believer in determinism. In his mind, the only role of randomness was to describe incomplete information.

more likely to come up the way it started than the other way, the probabilities being about 51% and 49% instead of the ideal 50–50. This deviation from true randomness is not that great, and the coin toss can safely still be used to decide who gets the kick-off in a soccer match (and the referee can start by shaking the coin between his hands so that nobody knows how it starts).

Another area of interest to Diaconis has been card shuffling, where he has asked and answered the question of how many times a deck needs to be shuffled before it becomes random (needless to say, he can shuffle a deck over and over without it ever becoming random). When you buy a new deck of cards, it comes with all suits ordered one after another. You shuffle a few times and spread the cards out on the table. If you see a sequence of five consecutive hearts and another of seven consecutive clubs, you would probably conclude that you have not shuffled enough and the deck is not yet random. In order to treat this as a probability problem, it must be determined how to describe shuffling in mathematical terms and it must be decided what it means that a deck is random. Let us not go into these details, but just think of your favorite way to shuffle cards and whatever intuition you have for randomness. Diaconis' results are very interesting. He showed that with less than five shuffles, the deck is still essentially nonrandom, but after seven, it is very random. The change from nonrandom to random occurs very rapidly between shuffles five and seven, a card shuffling ketchup bottle effect.

Diaconis has also been interested in the randomness of dice but has concluded that this is far more difficult than coin tosses or card shuffles. Not too surprising. After all, Sir Francis Galton already knew that dice hurtle variously and tumble wildly about.

FINAL WORD

In Chapter 1, we started our journey into the world of probabilities by talking a lot about coin tosses as examples of randomness. Our journey has taken us to many places, and you have seen how probabilities enter our lives daily. We have talked about court trials, medical trials, casinos, opinion polls, German tanks, and Mendel's wrinkled peas; you have learned about random walks, expected values, margins of error, and even something called Chebyshev's inequality. With Persi Diaconis' stroboscopic investigations into the randomness of a real coin toss, I feel that we have come full circle, back to where we once started, to the pure and simple phenomenon of a coin toss. This is therefore a good place for me to bid farewell and leave you on your own in the world of probabilities. Hopefully you have learned to understand them better and perhaps even view them as your friends. They deserve it, those little numbers that rule our lives.

Index